AMAZING PIPELINE STORIES

AMAZING PIPELINE STORIES

How Building the
Trans-Alaska Pipeline
Transformed Life in
America's Last Frontier

By Dermot Cole

Epicenter Press
Fairbanks/Seattle

ABOUT THE COVER PHOTOS
Front: Welders relax in a section of pipe (Fairbanks Daily News-Miner).
Inset: Maggie Joyner works as a "bus swamper" in Valdez (Neal Menschel).
Back: U.S. Senator Mike Gravel celebrates atop the pipeline (Fairbanks Daily News-Miner).

Epicenter Press Inc. is a regional press founded in Alaska whose interests include but are not limited to the arts, history, environment, and diverse cultures and lifestyles of the North Pacific and high latitudes. We seek both the traditional and innovative in publishing nonfiction tradebooks, contemporary art and photography giftbooks, and destination travel guides emphasizing Alaska, Washington, Oregon, and California.

Publisher: Kent Sturgis

Editor: Tricia Brown

Proofreader: Lois Kelly

Mapmaker: L.W. Nelson

Cover Design: Leandra Jones

Text design: Newman Design/Illustration

Printing: Transcontinental Printing

Library of Congress Catalog Card No. 97-060785

To order single copies of Amazing Pipeline Stories, mail $14.95 (Washington residents add $1.29 sales tax) plus $5 for first-class mailing to: Epicenter Press, Box 82368, Kenmore, WA 98028.

Booksellers: Retail discounts are available from our distributor, Graphic Arts Center Publishing, Box 10306, Portland, OR 97210, Phone 800-452-3032.

PRINTED IN CANADA

First printing May 1997

10 9 8 7 6 5

To my wonderful wife, Debbie, and to our children, Connor Patrick Cole, Aileen Elizabeth Cole, and Anne Susan Costello Cole, this book is dedicated. They put up with me while I was writing it.

Acknowledgments

In writing this book I have relied on written accounts from the pipeline period in newspapers and magazines throughout the United States; numerous books; tape-recorded interviews conducted during construction as part of a private sociological study, and later as part of a state-funded history project to document the pipeline; interviews with people who were in Alaska at the time; and a host of government reports and other documents.

In particular I would like to mention the writing and research of the following journalists in Anchorage and Fairbanks: Howard Weaver, Bob Porterfield, Jim Babb, Sally Jones, Rosemary Shinohara, Richard Fineberg, Craig Smith, Tom Snapp, and Jane Pender.

Other writers or researchers whose work has been invaluable are: Robert Douglas Mead, Mim Dixon, Allen Chesterfield, Potter Wickware, Jack Kruse, Joe LaRocca, Sue Fison, Mike Goodman, William Endicott, and David McCracken.

I would like to thank Gretchen Lake and the rest of the staff at the University of Alaska Archives for helping make my hours in the bowels of the Rasmuson Library productive. The Larry Carpenter collection, donated to the library after his death, contains a wealth of material about the pipeline, as do other archival collections.

I would like to thank Kelly Bostian, managing editor of the *Fairbanks Daily News-Miner*, for his cooperation and support during this project. I have worked at the *News-Miner* for most of the past twenty years as a reporter, editor, and daily columnist.

I cannot say enough for the efforts of Kent Sturgis of Epicenter Press and editor Tricia Brown, whose editing skill has added much to this book.

I would also like to thank: Debbie Carter, Terrence Cole, Pat Cole, John Hewitt, Mary Fenno, Michael Carey, Elstun Lauesen, Paul Helmar, Steve Porten, Dave Haugen, Ben Logan, Heather Kendall, Mike Davis, Dirk Tordoff, Jan Wilson, Ray Dexter, and Sunny Carpenter.

CONTENTS

FOREWORD
What Happens When It Goes Boom   11

INTRODUCTION
The Thin Silver Line   15

SECTION 1
The Work Force
    Hurry Up and Wait   24
    Of Unions and Reunions   32
    The Big Man in the Floppy Hat   33
    Trucking on the Kamikaze Trail   38
    Women at Work   45
    The Welders from Tulsa   57

SECTION II
Skinny City
    The Camps and the Pump Stations   67
    An Eight-Hundred-Mile Neighborhood   71
    Henry Kissinger's Dream   85
    Camp Rules   89
    All You Can Eat   91
    The Naked Truth   99
    New Beginnings in the Pipeline Parish   100
    Tune In Next Time   103
    Letters Home   106
    Farthest North Gator   107
    A Piece of the Pipeline   109
    Patriotic Pipeliners   112
    Muleskinner on the Haul Road   114

SECTION III
Sudden and Unexpected Wealth
    Sudden and Unexpected Wealth   118

SECTION IV
Power and Vice

Looking for Live Ones   132
The North Star Terminals Murders   142
Theft and Waste   146

SECTION V
Growing Pains

An Occupied City   152
In the Path of the Pipe   162
Squeezing In   167
Road Warrior   174
From Second Avenue to Two Street   176
That Shade of Yellow   186
The Busy Sound of Success   190
Plain Greed   196

SECTION VI
The Aftermath

Bigger and Busier   198

Almanac

By the Numbers   206
Pipeline Cracks   209
Pipeline Timeline   215

Sources   219

Index   222

About the Author   224

Steve McCutcheon/Alyeska Pipeline Service Company

On June 20, 1977, the first oil flowed from Pump Station 1, the starting line. Posing beside a "pig launcher" are William J. Darch, left, then president of Alyeska Pipeline Service Company, and Derrick Dunn, Alyeska Startup Commissionary Manager. Mechanical plugs, known as pigs, would move through the pipeline to improve the flow of oil or aid in inspections.

FOREWORD

What Happens When It Goes Boom

On a sunny June day in 1977, the day the Trans-Alaska Pipeline went into operation, the city editor of the *Fairbanks Daily News-Miner* assigned me to write about some of the amazing things that had happened in Fairbanks during the building of the pipeline.

That day I wrote an article about the two-bedroom rooming house that forty-four people called home during the housing crisis; the transformation of Second Avenue into "Two Street," where anything could be had for a price; the pipeliners who favored pointy-toed boots and cowboy hats; and how the endless lines turned many a twenty-minute errand into a two-hour excursion.

The next morning, the city editor sent me downtown to sample public opinion about this momentous event, the startup of the pipeline. The flow of oil was limited to a one-mile-an-hour crawl as it left Prudhoe Bay during the pipeline shakedown phase.

I grabbed my notebook and walked down Second Avenue, still a tightly packed collection of bars, to the Co-Op Drug Store, which was holding on to its reputation as the place where "everyone meets everyone."

It had been nine years since the discovery of the largest oil field in North America. After the expenditure of $8 billion on a three-year crash program that was routinely described as an "invasion" in Fairbanks, I asked people what they thought about it all.

A retired carpenter said OPEC would have a harder time jacking up oil prices. A heavy-equipment operator said the state might stay out of debt, while a mechanic saw an end to the big-money jobs. Most people didn't seem very excited except for a secretary to a top pipeline official, who said she was proud to be a part of the whole thing.

At the time, it was my impression that people in Fairbanks

seemed emotionally drained when it came to talking about the pipeline. A cab driver named Bernie explained all.

"It's a lot like having a baby," Bernie said authoritatively. "You've got a long time to think about it. It really doesn't amount to much."

This comment by the cabbie, after it appeared in print with a big headline that said "Pipeline's like 'having a baby,'" led to a complaint I've never forgotten. A woman wrote a letter to the editor saying it was obvious how many times Bernie had given birth.

"It's even more obvious how many babies Dermot Cole has 'had,'" she wrote. "As a more reliable source, I can assure you (and Bernie and Dermot) that 'having a baby' is exciting, awakening, exhilarating, amazing, beautiful, astounding—in short anything at all except anti-climatic."

She was right, of course. And looking back at the construction of the Trans-Alaska Pipeline, though it may not have been apparent to Fairbanksans weary of rapid change in 1977, it surely was "anything at all except anti-climatic."

Amazing Pipeline Stories is about the labor pains and other events that preceded the delivery of Alaska oil in the biggest boom Alaska has ever known.

The late Frank Moolin, Jr., the senior project manager for the pipeline, liked to say he practiced "management by exception." He didn't want to hear about anything that was going right, he only wanted to hear about the exceptions.

This book includes a lot of the exceptions, at least in the way that the term might apply to life in the cities and pipeline camps. The exceptions are the things that grab our attention and make for compelling stories.

While the pipeline was under construction, most Alaskans managed to send their kids to school, go to church, and take the dog for a walk. And yet newspapers outside the state portrayed Fairbanks, and to a lesser degree Anchorage, as Gomorrahs of the Far North during the pipeline boom. The conditions described in such articles as the *New York Times Magazine* piece that carried the headline "Blood, toil, tears and oil: How boomtown greed is changing Alaska," all existed, but they were by no means universal.

At one of the many complaint sessions among Fairbanks business leaders about why reporters always seemed to focus on what was wrong with the pipeline, Larry Holmstrom, a talented and media-savvy aide to the governor, passed on the observation that "the house that doesn't burn isn't news."

The house that doesn't burn is not normally a major part of journalism or history, but its existence should be noted nonetheless.

I lived not far from downtown Fairbanks during the pipeline boom, and I enjoyed the experience. Of course, I was in my early twenties, and I wasn't married and didn't have any kids, factors that may have had something to do with it.

I joined the legions moving to Alaska in 1974 when I was twenty-one, becoming one of the Outsiders who were putting a steady strain on the traditional Fairbanks quotient of elbow room. I didn't come to work on the pipeline, but to visit two of my brothers and one of my sisters, who had moved up to attend the University of Alaska. Pipeline impact and all, I liked it so much I decided not to go back to college in Montana.

I moved onto the couch in an old cabin in downtown Fairbanks that my sister had rented, one in which the rent was not raised, and stayed in Fairbanks to earn a degree in journalism. It was an exciting time to be there, especially for someone who didn't know what it was like before and didn't mind the pipeline commotion.

The peak of the boom occurred in 1975-76 and began to taper off the following year. In the article I wrote on the day the pipeline went into operation, I quoted a Fairbanksan who relayed a perception that seemed typical: "I had real fears at the peak of the activity—to sell out and move to New England because I thought the pace and the atmosphere of, say, a year or two ago, would continue. Thank God I decided to stay, and thank God I was wrong, and thank God it's over—at least for a while."

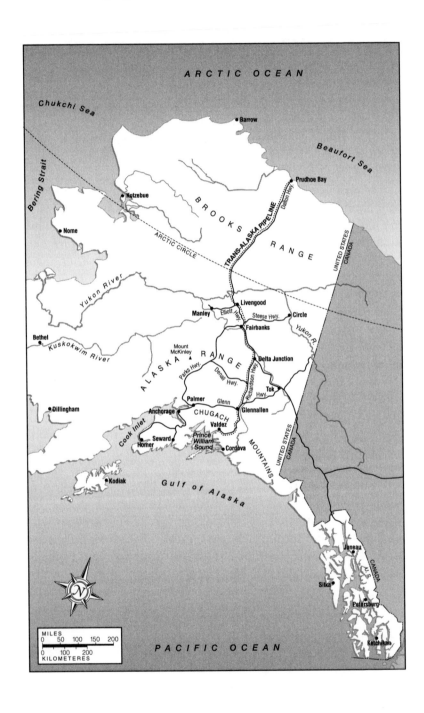

INTRODUCTION

THE THIN SILVER LINE

The ancient traveler Philo of Byzantium wrote of the Seven Wonders of the World in the third century B.C.: "If a man goes to the different locations, sees them once and goes away, he immediately forgets. The details of the works are not recalled, and memories of the individual features fail. But if a man investigates in verbal form the things to wonder at and the execution of their construction, and if he contemplates, as though looking at a mirror image, the whole skillful work, he keeps the impressions of each picture indelible in his mind. The reason for this is that he has seen amazing things with his mind."

Though all of the Seven Wonders but the Great Pyramid are now known only through archaeology, they gained fame as the most amazing engineering achievements of ancient times.

Evoking the imagery of Philo of Byzantium, the American Society of Civil Engineers in 1994 compiled a list of the "Seven Wonders of the United States."

Rather than tombs, statues, gardens, and temples, the engineers selected these nineteenth- and twentieth-century wonders: the Kennedy Space Center in Florida, the launch site for the Apollo moon project; Hoover Dam in Nevada, built with enough concrete to pave a highway from coast to coast; the twin 110-story towers of the World Trade Center, which can house 50,000 employees; the 42,696 miles of the Interstate Highway System; the elegant Golden Gate Bridge in San Francisco; the Panama Canal, selected because it was made possible by American ingenuity and money; and the Trans-Alaska Pipeline.

The successful design and construction of the eight-hundred-mile pipeline from Prudhoe Bay to Valdez, the engineers' society said, "stands as one of history's most difficult engineering feats."

The nation's largest pipeline crosses three mountain

ranges, the Yukon and thirty-three other rivers, and eight hundred streams. It bisects Alaska from Prudhoe Bay on the North Slope, to Valdez on Prince William Sound, carrying about 10 percent of the oil consumed every day in the United States.

Built in three years and two months, the pipeline employed a peak work force of 28,072 and did more to shape the development of Alaska than all the gold rushes combined.

The Alyeska Pipeline Service Company, the consortium of seven oil companies that built the pipeline, placed a priority on speed. The contractors in the five pipeline sections, nicknamed Slackloop, Hotdope, Snoopy, Sidebend, and Rocsaw, were all players in the weekly horse race known as the "Alyeska Sweepstakes," the winner of which was the one that finished the most work.

Alyeska tracked its progress on the calendar because every day the oil stayed in the ground was another day before the oil companies could start selling oil and collecting on their investment.

The demands of the pipeline timetable shaped the events of the construction years. One senior pipeline official complained about an attitude of "gold-plating" on the project, which he said led to the idea that "only the best and the most is good enough."

Not since World War II and the haste to fortify Alaska against the threat of Japanese attack had a development plan been mobilized with such fury in the far north.

Alyeska President Ed Patton, a veteran Exxon official who had served in the Navy in World War II, used analogies to the Normandy invasion and the Battle of Midway to describe what was first known as the Trans-Alaska Pipeline System, or TAPS.

The critics of TAPS, Patton said, were like those who complained about the loss of the carrier *Yorktown* in the Battle of Midway, while failing to mention that the Allied victory there turned the course of the war in the Pacific theater.

"Just like TAPS, the Battle of Midway did not go perfectly—but that does not change the fact that Midway, like TAPS, was a resounding success," Patton said.

Many environmental critics who questioned the justification for building the pipeline in the first place remained steadfast in their opposition from start to finish. They viewed the pipeline, not as a technological marvel or as a victory over harsh conditions, but as a pernicious influence that had intruded into and was destroying America's last wilderness.

The head of the Sierra Club Legal Defense Fund said that when the oil ran out in the distant future, the pipeline would be removed, as the oil companies had agreed, but the road built alongside it would be "as permanent as the pyramids."

To prevent the ground from thawing in areas where permafrost is present in the soil, two-inch heat pipes with aluminum radiators were placed atop the vertical supports. The pipes, which contain anhydrous ammonia, assist in transferring heat, to keep the soil frozen.

David Brower of the Friends of the Earth spoke for many when he branded the pipeline as the "greatest environmental disaster of our time." Other environmentalists said that the years of delay between the 1968 discovery and construction, years of lawsuits and public hearings, at least forced the oil companies to make the line safer than it would have been otherwise.

The biggest accident occurred on Good Friday in 1989, when the *Exxon Valdez* ran aground on Bligh Reef in Prince William Sound, spilling 240,000 barrels of oil and igniting a bitter environmental controversy that raised echoes of the original knockdown fight to build the pipeline.

While there are critics who have continued to question Alyeska's performance, the company does enjoy a better public image than it did during the intense spotlight of the construction years, when the quiet wilderness became a noisy landscape, crowded with an invading army of construction equipment, Texas pipeliners, and yellow Alyeska pickups.

During its first two decades, the highly profitable pipeline carried about twelve billion barrels of oil to tankers in Valdez. Estimates of annual oil company earnings from the North Slope fields ranged from $2 billion to as much as $4 billion, and the pipeline operated more than 99 percent of the time, pumping enough oil to fill about seventy tankers a month.

It all began with a turning point in Alaska's history, the 1968 oil discovery by Atlantic Richfield and Humble on state land on the North Slope. A geologist said the natural gas escaping from the well sounded like four jets flying overhead.

The daily oil flow from the discovery well, Prudhoe Bay State No. 1, would have been enough, one writer said, to run a car for nearly two centuries. The well's daily gas flow would have been enough to cook all of a family's meals for two thousand years.

A confirmation well, which tapped into the same massive formation nearly two miles below ground and seven miles away, helped establish the enormous dimensions of the strike. It turned out to be the largest oil find ever in North America and the eighteenth largest oil field in the world.

To many in Alaska, the news brought visions of a glorious financial future befitting the optimism of Governor Wally

Alyeska Pipeline Service Company

A section of pipe is prepared for burial in the tundra about forty miles south of Prudhoe Bay. By mid-August 1975, more than a hundred miles of pipe had been installed.

Hickel, whose attitude was aptly summed up when he said that the address of Arctic Alaska should be "No. 1 Wall Street for Alaska's Future."

Pipeline proponents compared the pipeline to a line across a football field, a thread across a newspaper, a string across a basketball floor, a scratch across the back of the hand, and a strand of spaghetti a million times longer than it was wide. They repeatedly pointed out that the pipeline would cover only twelve square miles of Alaska when complete, as if that somehow cut the project down to size.

Pipeline opponents compared the pipeline to a scratch across the Mona Lisa, or as another critic said, a narrow line "as fatal to wildness as a sliver through the heart."

Oil historian Daniel Yergin has noted that the North Slope of Alaska was unlike any other place where oil had been developed and the technology did not exist to produce it at the start.

"There were no roads across the tundra, and beneath was the permafrost, that part of the soil that was permanently frozen. Normal steel piling would crumble like soda straws when driven into the permafrost," Yergin wrote.

Ideas on how to get the oil to market ranged from a proposal by General Dynamics to build a fleet of six nuclear-powered submarine tankers, to studies by Boeing of immense twelve-engine jet tankers. There were also the experimental voyages of the *S.S. Manhattan*, an ice-breaking tanker that traveled the Northwest Passage.

From early in 1969, however, the oil companies had placed their bets on a $900 million pipeline across Alaska, to be completed by 1972. The oil companies, as well as President Nixon and the state of Alaska, resisted demands by environmentalists to look at a pipeline route that would have gone through northeast Alaska and south through Canada. That route would have eliminated the need for a vast tanker operation from Valdez.

Determined to go across Alaska, the oil companies ordered $100 million worth of forty-eight-inch diameter pipe from Japan, which turned out to be one of the few real bargains in the entire project.

The pipe would remain in storage yards at Valdez, Fairbanks, and Prudhoe Bay for five years, however, as the unresolved issues of Native land claims and environmental lawsuits froze the project. Congress approved the Alaska Native Land Claims Settlement Act in 1971. Meanwhile, the environmental debate over the pipeline tested the new law that established the Environmental Protection Agency. The court challenges continued until 1973, when Congress intervened. Casting a vote in the U.S. Senate for only the second time,

Vice President Spiro Agnew broke a 49-49 tie on an amendment by Alaska Senator Mike Gravel that limited further court delays.

The President signed the act authorizing the pipeline that fall, just as the Arab oil embargo created gas shortages in the United States. Motorists had to wait for hours in long lines to buy gas, if the stations had any at all. Gasoline prices rose by 40 percent, and many Americans learned for the first time that their big cars from Detroit were running on oil imported from the Middle East.

Chancy Croft, who was president of the state senate during pipeline construction, pointed out that the pipeline became economically feasible because of the Arab oil embargo and environmentally workable because of the U.S. Geological Survey.

"It was the U.S. government and not Alyeska who forced substantial revision of the original design criteria," Croft wrote. "The federal government set the standards to which the pipeline was built."

Croft also said that because of Alaska's unions, the oil industry had to "take into account the standard of living of the workers. No more cheap wages and lousy working conditions."

With overruns and changes in design, the cost of the pipeline climbed like a rocket, eventually topping off at $8 billion.

In his book about the construction, *Journeys Down the Line*, author Robert Douglas Mead tried to put the cost into perspective. In a calculation he made when the official estimate was $7 billion, Mead said if you spent $1,000 a day it would take 19,178 years to go through $7 billion.

The task of the pipeline builders was so complex, Senior Project Manager Frank Moolin, Jr., once said, "It's impossible even for a hundred people to fully comprehend what's going on in the project."

In 1976, a Fairbanks anthropologist observed that stories about the pipeline had not become part of local folklore and traditions, unlike the Gold Rush, the building of the Alaska Highway, and the Distant Early Warning line. Commenting on this phenomenon, newspaper columnist Ben Harding said that because of the "incredible waste and inefficiency," the pipeline just wasn't held in the same high regard.

In time though, the stories about the pipeline, including many about waste and inefficiency, did become part of the folklore of Alaska.

People who were in Alaska during the pipeline boom haven't forgotten the Oklahoma welders who rioted when they couldn't get steaks for lunch, the experience of life in the pipeline camps, the truck drivers who earned more in a year than the vice president of the United States, or the charged atmosphere of Second Avenue in Fairbanks and Fourth Avenue in Anchorage.

Amazing Pipeline Stories is not a definitive history of pipeline construction, its problems and controversies, nor is it a treatise on the impact of oil on Alaska. There is room for others to follow and to thoroughly investigate the long-term implications of oil development in Alaska.

Rather, what follows is a portrait of the most tumultuous period in Alaska during the twentieth century, a glimpse at the boom that overshadowed all others.

SECTION I

THE WORK FORCE

Fairbanks Daily News-Miner

Waiting became a necessary pastime for workers when a decision was coming, or they lacked tools or supplies, or parts were on the way. Here a pair of welders take a break in the pipe.

"At the time we arrived there, it looked like Atlanta after the Civil War. There were bodies everywhere—asleep. A foreman woke up, and went over, and began waking people up who—it took about five minutes or so, but they finally got up, got the machines going, and began drilling operations at that time. But, for God's sake, when the governor is coming to take a look at the pipeline project, you'd think they'd at least be up and looking busy, if not doing something."

— **State Pipeline Coordinator Chuck Champion describing visiting a pipeline work site with Governor Jay Hammond.**

HURRY UP AND WAIT

A small photograph on page 20 of the June 2, 1975, issue of *Time* magazine changed Al Fleming's life. At the time he was a thirty-eight-year-old teacher with a master's degree, who was earning $10,600 a year in Utah.

When that week's issue of *Time* reached his mailbox, he read the article headlined "Rush for Riches on the Great Pipeline."

"They come by jets screaming in from Houston and New Orleans, or in mud-covered Winnebago trailers swaying up the Alaska Highway. They come in ancient station wagons, the kids frisking in back, the husband hunched over the wheel and the exhausted wife dozing fitfully in the front seat. They are the latest breed to head for Alaska with the burning desire to strike it rich."

The article quoted Fairbanks police Captain Lewis Gibson as advising against looking for work in Alaska, saying, "I can't imagine a worse place in the world to be than Fairbanks this summer." But Fleming couldn't resist.

What would stick in his mind forever was the photograph that showed a pipeline welder proudly holding his paycheck up for the camera. For an eighty-four-hour week, the welder earned more than $1,300. The caption read, "Pipeline welder displaying fat weekly check."

Fleming painted enough houses that summer for a one-way ticket to Fairbanks. He gave himself six weeks to get a job, or give it up and go home to his wife, four kids, and the next school year. He was persistent and lucky, landing a job as a Teamster warehouseman in Fairbanks.

Fleming saved every check and sent the money home to his family, which gave him a new start in Alaska. Like many others who remember the pipeline life, he says it all seems a bit unreal.

After the pipeline he returned to teaching school, his first love, only this time he did it in Fairbanks. In 1993, the veteran educator was named the city's Teacher of the Year. He retired the following year.

"The pipeline was like Disneyland, Fairyland, and Mother Goose Land all wrapped into one," he said. "What had surprised me was, I saw these guys with these fantastic jobs. These guys with these fantastic salaries would do everything they could to screw themselves up. I didn't understand it. I felt I was the luckiest guy in the world."

Fleming was one of thousands who migrated to Alaska in search of big money. Others heard of the rush for riches in union halls or listened to tales from friends or relatives who had struck it big. They came from all walks of life, ranging from former Milwaukee Braves infielder Johnny Logan and former Algerian lightweight boxing champion Hocine Khalfi, to the four King brothers, professional pipeliners from Bald Knob, Arkansas.

Many people joined the parade of job-seekers after reading a newspaper article or two, ignoring the advice of those who warned against going north without a job. After a *New York Daily News* article about the pipeline in early 1975, Alyeska's office in Anchorage received 6,576 letters and 1,370 phone calls in one month.

The *National Enquirer* ran one of the countless ads about how to get jobs on the Alaska Pipeline: "$1,200 weekly, Alaska pipeline jobs? Guaranteed 'real facts.' Rush $1" to an address in West Palm Beach, Florida.

Anyone sending $1 to the address got a response back

saying that the "real facts" about the pipeline jobs were that none were available and not to bother going.

Another ad said getting a pipeline job didn't require experience, "pull," or youth. All it required was the $3.95 "Alaska Pipeline Information Package."

With or without the information package, thousands of people who came to Alaska looking for pipeline work were disappointed at how tough it was to get hired.

John O'Ryan, a reporter for the *Seattle Post-Intelligencer*, was given $100, a tent, a backpack, and a plane ticket to join the great stampede. O'Ryan was no stranger to Fairbanks, having been the managing editor of the newspaper there in the 1950s.

On his pipeline job hunting trip, he spent his first night on the floor of the airport because there were no rooms to be had. The next day he made the rounds of the union halls and learned about the job lists.

Regular members of the Laborers Union, for instance, who had worked more than eight hundred hours in the previous year were on the "A" list, which gave them first crack at any jobs. Members with more than a hundred hours, but less than eight hundred hours, were on the "B" list. They would get jobs after the "A" list members got the jobs they wanted.

The "C" list was home to people from Outside with two years of experience and to Alaskans with no previous connection to the union. Or to those who could fake those qualifications, with phony documents. O'Ryan learned that some boarding houses had been dispensing phony back-dated rent receipts, but the unions had ruled out accepting them as proof of residency.

One step down in the pecking order were the people on the "D" list, who rarely got jobs. Had O'Ryan stayed in Alaska, he would have added his name to the "D" list.

The requirements differed somewhat from union to union, but O'Ryan found that he would have to wait at least a year for a job. He hitchhiked to Valdez, spent all of his $100, and had to wire the newspaper to send him a plane ticket out.

Getting union jobs was a straightforward matter to some people who put their names on the list and waited their time, but others found the process unfair.

Charlie Backus/Alaska State Library
In January 1975, picketers marched at the Culinary Workers' Hall in Fairbanks.

"It seemed to get a union job, one needed, in this order: a good Southern drawl, cowboy boots, and/or a thousand dollar bill. A friend with a yellow pickup truck with 'Alyeska Pipeline Service Company' on the side wasn't exactly a hindrance either," said Dave Otnes of Petersburg.

After getting into the system and landing a spot on the "A" list of the Laborers, Teamsters, or other union, many pipeline workers didn't hesitate to quit during the peak of the project. That's because they could get hired back at any time. The turnover in some areas was 5 percent a week.

"Some of those guys were just tourists disguised as workers," said Alyeska President Ed Patton.

Just as for the fortune seekers from Outside, pipeline work presented a great opportunity for Alaskans, both those with considerable experience in construction, as well as students and others.

At 11:00 a.m. on January 1, 1974, Thomas Swarz started forming a line outside the union hall of Laborer's Union Local 942 in Fairbanks. He had walked six miles in the near-zero weather, bringing his sleeping bag so he could be first in line when the union office opened the next day. More than a hundred people spent that winter night camped outside the union office to get their names on the list.

The children of many prominent Alaskans opted for a shot at pipeline work. Four children of Alaska's U.S. Senator Ted Stevens worked on the pipeline: two Teamsters, one pipeliner, and one laborer.

"It's amazing for me to see the change in them. Here they are working six and seven days a week and ten or twelve hours a day when back home I couldn't get their attention on the dishes for more than thirty minutes," Stevens said.

While there were people who earned their money on the pipeline, there were also those who said that because of the surplus of workers on any given crew, they had never had so much time on their hands. Publicity about idle workers and tales of crews sleeping on buses became the stuff of legend and much speculation.

One of the jobs Dan Gross had was to drive a crew out of Franklin Bluffs to the work site and drive them back at night.

"In between, sleep, eat, read and write. I didn't know it at the time, but I was about to discover what the Trans-Alaska Pipeline was really all about—boredom," he said.

He wrote that the bus he drove regularly resembled: "(1) A dentist's waiting room. (2) Company HQ on a Siberian battle-field. (3) A small PX. (4) All of the above."

The answer was No. 4.

It was often said that because of the "cost plus" contracts, the pipeline contractors had no incentive to keep their work crews trimmed down to the lowest possible number because all labor expenses were reimbursed by Alyeska.

The contracts were called "cost plus" because the contractors did not have to decide how much they could afford to expend on labor. They just submitted the bills to Alyeska.

Alyeska officials said that the contractors did have an incentive to hold down costs, which was to preserve their reputations so they could get work in the future. How much of this attitude filtered down to the lower parts of the organization was debatable, because pipeline workers routinely said that there were always more than enough workers assigned to most jobs.

Pipefitter Potter Wickware said it was frequent, "if not

usual, for crews to have twice as many men as they need to do the job."

"You think I'm talking about goldbricks, but I'm not," he said. "I'm talking about your standard working man, a person like me. Here we are in this strange situation. They pay us like princes, and they don't give us any work."

In a report trying to document what the state claimed were excessive pipeline costs, former Watergate prosecutor Terry Lenzner alleged that excessive idleness was caused by bad management and added hundreds of millions of dollars to the cost of the pipeline. Lenzner, hired by the state of Alaska to investigate pipeline costs, claimed that Alyeska could have built the pipeline for $1.5 billion less than it did. The state was interested because the construction costs would be factored into the pipeline tariff and drive down the so-called wellhead price of oil on which state taxes and royalties were based.

Lenzner said that "much of the so-called loafing" was because crews didn't have work to do, they were sent out in bad weather, they were awaiting engineering decisions, or they hadn't been told what to do.

"A common abuse, for example, was workers sitting in buses for hours or even days at a time while drawing full salaries," Lenzner said.

He quoted a letter from a welder to the president of Alyeska in which the worker complained about a day on which he said he would be paid for ten hours, when there were only two hours of work. "My father taught me to give a man a day's work for a day's pay," T. Sherwin wrote. "I'm asking you to let me work for my money."

A report to the owner companies of the project said that men sitting in buses when they should have been working added up to losses of thousands of hours a week.

In a memo titled "Alyeska Labor Excesses," an Arco official said in early 1976: "I am shocked and dismayed, that never have so many done so little for so much; dismayed that these abuses had not come to light sooner."

Another Arco official said it wasn't the workers' fault.

"Many of the abuses that have been charged to the unions, such as too many men on the job and doing nothing, or sitting in buses playing cards, or not working because of weather, buses delaying to get to camps or work sites, or stopping at bars before reaching the camp site, or performing work in a sloppy manner, or the large number of welds that are rejected without remedial action taken, are principally the result of the contractors' failure to exercise their managerial supervision and enforcing the necessary contract positions with the workers," said V.R. D'Alessandro of Arco.

Alyeska's Patton, in an interview on the CBS news program *60 Minutes*, said that in none of the allegations about idle workers did he find any evidence of poor management. "I find a number of things in this project as evidence of a decreasing state of conscientiousness on the part of Americans in general," he said. "I mean, you see it in the attitude of the kids in college, you know, the rioting in the late '60s, and that type of thing, the willingness to accept welfare and food stamps."

Others said that the rigid jurisdictional limits by the unions, the source of almost daily disputes, also had a lot to do with idle time.

Teamster leader Jesse Carr didn't agree. Asked once to rate Alyeska's management efficiency on a scale of one to ten, Carr said it was off the scale. "I'd put 'em minus five. You know, in all the years that I've been in the construction business and tried to run a local union, I have never seen anything so mismanaged in my twenty-five years in Alaska."

No matter if it was goldbricking or a perfectly rational break between tasks, the workers sleeping in buses created real problems when they were seen by visiting VIPs.

In one such incident the president of Exxon was touring the pipeline when he saw a bus being towed by a side boom tractor with the bus driver asleep in the bus. As with other bus drivers, this one was free to sleep during the day because he had no responsibilities but to drive the bus back to camp that night.

Senior Project Manager Frank Moolin, Jr., said that in a project so immense, there were bound to be times when there wasn't much work to do because of the complicated logistics.

"Idle time is inherent in any construction operation," Moolin said. "If you spend any time watching the construction of a high-rise building, you will find that the workers don't have anything to do a substantial portion of the time."

Pipeline managers said that the situation they faced was like the Army, with a good deal of "hurry up and wait" time because many actions depended on a precise sequence of events. A lot of factors could combine to screw up the schedule, from government paperwork to the weather.

"Frequently during any normal work day, a worker inevitably will have nothing to do," said Moolin. "Workers on a pipeline project often work like mad for a few hours, complete their tasks, and then wait for several hours until the next job in the sequence is completed."

Moolin spent thirty to forty hours a week flying along the pipeline, and he said he checked on every instance he could to see if workers on buses were supposed to be there or not. He boarded any bus he saw parked with workers on it to find out why they weren't working.

In the best book written about the building of the pipeline, *Journeys Down the Line*, author Robert Douglas Mead concludes that low productivity of the work force was "the natural product of the two bureaucracies, Alyeska's and the unions', working against each other."

The unions supplied inexperienced workers. There was very high turnover and jurisdictional rigidity that led to bus drivers sitting idle ten hours a day. On the other hand, he said, the pipeline suffered from poor scheduling with crews sent to sites where there was no work.

"It is not necessary to cast moral stones at the men or to assume that many were there merely to draw pay and avoid work," Mead wrote. "A good deal has been gossiped to this effect in newspapers and magazines, more recently in books, but that was not what I saw; it is the view of some who signed on in order to write about the work, not do it."

OF UNIONS AND REUNIONS

It was a Generous family.

Mabel A. Generous, a senior clerk typist for Alyeska in Fairbanks, counted the most relatives working on the pipeline. Sixteen people in her family put in time on the line. They included: brothers Francis Esmailka, Peter Esmailka and Berchman Esmailka, Jr.; many first cousins: Adolph Ekada, Lucy Peter, Robert Smith, Michael Smith, Walter Stickman, James Demoski, Peter Demoski, John Demoski, Ida Hildebrand; and uncles: Peter W. Demoski, Rudolph Esmailka, and Austin Esmailka.

Other clans with large numbers on the pipeline were the family of Corine Leyva, a secretary for ITT/Arctic Services in Fairbanks, who had thirteen close relatives on the project, and Terry Flores Salgado, who had fourteen kin on the job.

Many father-and-son teams worked on the project, including Don Schouweiler and his son, Dennis, who were both powerhouse operators at Old Man Camp. Dennis's brother, Harold, was a heavy-duty mechanic at Old Man, and their brother Don, Jr., was a mechanic at Livengood.

"My sister Lorie is a senior at Petersburg High School," Dennis said in early 1976. "She's thinking about working on the pipeline. My mom doesn't want to work on the pipeline."

Another father-son team was that of Rod and Gil Perry. Rod worked as a laborer at Isabel Pass Camp, while his seventy-one-year-old father was a laborer at the pipeyard in Fairbanks. Rod Perry was a filmmaker who cast his dad in "Sourdough," a movie about an Alaskan sourdough pushed back into the wilderness to escape the influx of civilization in Alaska.

Then there was the King family of pipeline workers. Brothers Gary, Darrel, Kirk, and Bruce were all members of Local 798 from Bald Knob, Arkansas, one of the small towns made up almost entirely of pipeliners and their families.

They loved Harley motorcycles and had all started work in their teens as pipeliners. "I can't stand being inside," said Darrel. "This is my life."

THE BIG MAN
IN THE FLOPPY HAT

Instead of a construction hard-hat, one of the most important men on the pipeline construction project wore a floppy tweed hat that gave him the air of a laid-back farm hand.

"I wear the tweed hat I picked up in Ireland unless I'm exposed to falling objects," Frank Moolin, Jr., would explain. "Hard-hats are very uncomfortable in the weather we have."

The man in the floppy hat was a hard-nosed engineer who managed two hundred and sixty contracts worth more than $3 billion for the Alyeska Pipeline Service Company. He had an obsessive dedication to the pipeline that became legendary. Moolin worked fourteen hours a day, rarely took any time off, and took pride in being the first person to arrive in the office— and the last to leave. Three days a week he held full staff meetings at 6:30 a.m., and he was known to lock the door to discourage late arrivals.

Dave Haugen, manager of one of the southern sections of the pipeline, once got to the office at 4:00 a.m. just to be there before Moolin arrived. When Moolin arrived at 4:30 a.m., the two exchanged greetings without letting on that anything was out of the ordinary, but, Haugen said, "it really bothered him."

Haugen loved working for Moolin, who had a reputation, as one reporter put it, of "being so well-organized that he leafs through the *Harvard Business Review* at his tidy desk while he waits for the slow-moving managers he's about to fire."

"He was a classic leader in the best sense of the word," Haugen said. "I think that without question the pipeline would have not come close to the delivery date if it had not been for Frank Moolin. It was due primarily to his superhuman effort."

A 1956 magna cum laude engineering graduate of the University of Chicago, Moolin had worked on projects ranging from a refinery in Singapore to the Bay Area Rapid Transit System in San Francisco.

The 6-foot 3-inch, 215-pound Moolin, called "the barra-cuda" by some reporters, liked comparing the Trans-Alaska Pipeline to the Panama Canal and the Great Wall of China.

"Of course at that time [of the Great Wall] there were no labor unions to contend with," he said. "Environmentalists, I'm sure, were handled rather readily and rather speedily, with some summary form of justice. So I look upon this as a basic difference. Projects in the past were built on dictatorial types of conditions, and that's not what the future is all about."

Moolin became interested in the pipeline project when, in the early 1970s, the free flow of gasoline was stopped in America for the first time in decades, and the nation's motorists had to wait for hours to get filled up at the neighborhood gas stations. The national uproar over the gas lines helped speed Congressional approval of the trans-Alaska pipeline by changing the political climate. Under fire from constituents to do something about the energy crisis, congressmen found it easier to support oil development in far-off Alaska.

"I said to myself, 'What's more important than getting rid of the Arabs as the control of our destiny and as the control of our growth?'" The pipeline, he said, was "the biggest thing going."

ALYESKA SWEEPSTAKES

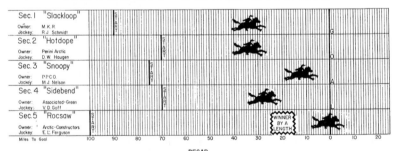

SECTION	1	2	3	4	5	TOTAL
1975 GOAL MILES OF PIPE	90	70	75	70	100	405
MILES INSTALLED TO DATE	60.7	41.1	67.4	47.1	103.3	319.6
MILES TO GOAL	29.3	28.9	7.6	22.9	N/A	85.4
MILES INSTALLED THIS WEEK	3.5	2.9	3.1	6.7	9.7	25.9

Period Thru _10/26/75_
Special Issue

University of Alaska Fairbanks Archives

Moolin's Alyeska Sweepstakes pitted contractors against each other in a race to finish their section.

Despite Moolin's leadership abilities on the pipeline, he was not everyone's idea of an ideal boss. Fear of the man was a big motivator for some Alyeska employees, who knew that in Moolin's view there were only two incentives—promotion or termination. He offered plenty of both.

"Guys used to feel bad when Moolin canned them. Now there have been so many that the stigma is gone, and it's like being sent to the bench by your coach," said one Alyeska employee.

With Moolin's stringent demands for performance, four of the five pipeline sections from Prudhoe Bay to Valdez had at least two managers within a year. The managers reporting to Moolin managed the contractors who handled pipeline construction, catering, equipment, and other services.

"Brown noses don't last long," Moolin would say. "They often can't keep up with the kind, quality, or quantity of work we have."

Moolin didn't like it when his managers went on R&R, and he had little patience for family matters that took them away from absolute commitment to the project.

"He demands total dedication, saying 'Your wife had a baby? So what?'" reported the *Engineering News Record,* a publication that named Moolin the construction industry's Man of the Year in 1976.

Joe Sledge, an Alyeska engineer, said that compliments from Moolin meant a lot. "You'll get your presentation back marked by his famous red pen, 'This is a really good show.' I know people who have saved every one because there just aren't that many."

He often showed up in the camps unannounced and flew by helicopter all up and down the line. A reporter said that Moolin could in all honesty say, "This morning, I personally observed on the workpad," and send the same message to camps hundreds of miles apart.

When coming upon workers asleep in buses, Moolin would walk in, wake the person who was slumbering, and hand him a business card on which he had written, "Have your supervisor call me."

A typical telex from Moolin might say: "Let me make my

position very clear. We are not making the progress we should be making. Last week we advanced only 1.8 percent."

To motivate his pipeline team, Moolin started what he called the "Alyeska Sweepstakes," in which the relative progress of the contractors working in the five sections of the pipeline was charted each week on productivity telexes.

The horse race results, published and widely disseminated within the industry, did not sit well with some of the contractor "horses," especially those at the back of the pack.

"He's used competition between sections effectively, but I told him this horse race business was getting counterproductive because it's frustrating," said David Perini, president of Perini Corporation, a key firm in one of five joint ventures. "People are facing things they've never been in before. Competition is okay, but you're criticizing people who are breaking their butts."

Early on in pipeline construction, the Bechtel Corporation was the management contractor, an arrangement that proved unwieldy. In what one historian describes as one of the "most humiliating defeats" in the history of Bechtel, the powerful construction firm was removed from its management role, a change pushed by Moolin. After taking over as project manager, he moved to shake up the operation and shift more responsibility to the field managers.

Moolin said he had learned to be less involved socially with his employees and that when terminations had to be made, they were quick. "I don't like lame duck situations," he said.

Some of the people he worked with figured that he was divorced or that he had never been married, because he worked all the time when he was in Alaska, and he didn't talk much about his private life. Truth was, his wife, Ruth, and their two children had stayed in New Jersey so his kids could get some measure of stability and finish school.

Ruth said she and her husband had moved nearly twenty times in twenty years, and the children had asked: "Everyone is from someplace. Where are we from?" They were at the point in their education where it seemed wisest to not uproot them from New Jersey, she said. While Moolin was in Alaska, he once took two one-day trips to New Jersey in one week to see his children graduate from high school and elementary school.

After the pipeline, Moolin created his own construction consulting company and served as chief executive officer of the Alaska International Industries conglomerate. He was a vice president of Western Airlines and actively involved in creating the hub system for that airline in Salt Lake City when he was stricken by leukemia. He died in 1982, shortly after his forty-eighth birthday.

In a deposition that became part of a mammoth case on pipeline tariffs assessed against Alyeska, Moolin, not long before his death, summed up his experience on the pipeline: "Perhaps I am being immodest, but I believe that in my career, I have accomplished many things. And foremost among them is the successful construction of the Trans-Alaska Pipeline System. I am confident that history will place the TAPS project among the highest technological achievements of engineering and construction. We had a very tough job to do, and we did it well."

Fairbanks Daily News-Miner

Wearing his trademark tweed hat, Senior Engineer Frank Moolin, Jr., escorted President Gerald Ford on a tour of the pipeline near Fairbanks in late 1975.

Fairbanks Daily News-Miner

Loading up with pipe in the Fairbanks pipeyard, December 1975.

"I wouldn't go through that winter of '74 for a million dollars
again. But I wouldn't give the memories up for $10 million.
The troubles we had. The way we helped each other out.
When you left Fairbanks, you didn't leave working for
Mukluk or Kaps . . . you left as a truck driver. If you met
another truck driver in trouble, you helped him. There was
no question, because if you didn't, he's liable to die."

— **North Slope Haul Road Trucker Larry Wegner, 1977**

TRUCKING ON
THE KAMIKAZE TRAIL

Songs about truck driving were a mainstay of country music in the 1970s, right up there with tales of cheating, drinking, and loving.

When country music star Freddie Fender appeared at Hering Auditorium in Fairbanks in 1976, the opening act featured truck driver and part-time country singer Sam Little performing his ode to the Haul Road. The song was *Trucking on the Kamikaze Trail.*

Little, a driver for Kaps Transport and a Teamster steward, knew first-hand about the hazards of hauling eighty-foot sections of pipe to points north of Fairbanks. "This song is a kind of a heroism song, but there is a little bit of heroism in trucking up here because it's a special kind of trucking," Little told one interviewer.

With guitar in hand, he sang:

> Now there's a road up north of Fairbanks,
> the Kamikaze Trail is well-known.
> It's a challenge to any truck driver,
> who really likes to carry their own.
> You know the haul is tough, the roads are rough,
> but the big rigs up and down her sail,
> Oh just a big happy family of gear-jammers
> a-trucking on the Kamikaze Trail.

The route that prompted the musical allusion to a suicide run was no superhighway, but it was an essential link in the building of the Trans-Alaska Pipeline.

The truck drivers followed the narrow and winding Elliott Highway to Livengood and then took the so-called TAPS road to the Yukon River, where they connected with the 360-mile Haul Road that had been built in five months in 1974 for $125 million.

Alyeska President Ed Patton, the man for whom the Yukon River bridge was later named, was fond of saying that by building the road to secondary highway standards Alyeska had made a great gift to the state, which took over the road in 1979.

"We could have gone over the Brooks Range by having to tow trucks over and let them down on the other side," Patton said. "We could have done that and it might have been cheaper."

When the road to the north came under heavy criticism after several truck accidents in the winter of 1974-75, the complaints centered not on the new road north of the river, but on the old road to Livengood, the Elliott Highway, which was steep and narrow.

Long-time Fairbanks educator Niilo Koponen commented at the time that the Elliott was a "nice Alaska road," which meant it was an old mining trail, not designed for trucks with extra-long loads of forty-eight-inch pipe. Highway Commissioner Walt Parker said the road was "the same old Elliott Highway" and that hazardous driving conditions "can only be eliminated by moving to a gentler climate."

Tow truck operator Rudy Voigt, who towed dozens of

trucks out of ditches along the Elliott, said that most of the accidents were caused by drivers who didn't know how to handle snow and ice. In some places the road was just wide enough that two rigs could pass. Voigt suggested that placing more road signs to warn of danger spots would help.

Trucker Phil Menges, a former grader operator for the state highway department, said the road did "amazingly well for how it was built."

"If anybody should be criticized for that portion of the road, it should be Alyeska because it's their pipeline, they're the ones getting material over it, and they're the ones that should foot the bill and straighten the road out, which eventually will be done. No doubt after most of the stuff is hauled, it will be a pretty good road," he said.

Jesse Carr, the Alaska Teamster leader who liked to say "Power is like a lady—if you have to say you are, you ain't," wasn't inclined to engage in long discussions about the need for improvements. On his orders, all truck traffic in the state was stopped for four days in February 1975.

The work stoppage was not called a strike, for strikes were prohibited by the contract with Alyeska, but "safety meetings" were something else. For four days Teamsters met twice a day to talk about safety. And the long haul trucks supplying the world's largest private construction project didn't move.

"The conditions are atrocious, and I'm not going to lose any men to that road," Carr told the press. "I searched my soul for quite awhile until I decided I'd had enough, and I shut down the road and made the state maintain it in a proper manner."

It was this action that Carr was referring to when he later told the *Los Angeles Times*, in a comment that was often repeated to show his power in Alaska: "I've got the hammer to shut it down, and I have."

In wielding the hammer, Carr was also seeking improved access by truckers to camps north of the Yukon for beds and meals and possibly to gain leverage in a dispute with a trucking company.

To keep the peace and get the trucks rolling again, better maintenance was promised by the state and Alyeska. In addition, the pipeline company agreed to be more hospitable to truckers at the camps.

Little, whose Citizens Band radio handle was "Fruit Tramp," said that the road was a tough challenge both before and after that brief shutdown. He said the toughest trips were with the loads of three sections of eighty-foot pipe.

"It was a good way for a truck driver with a good record to blow his reputation in one trip," said Little, adding that the pipe trailer used to transport the pipe "had a mind of its own."

"It wasn't just like pulling a trailer down the road," he said. "It turned on its own. You just had to be thinking ahead of it all the time. It's kind of like backing up a set of doubles, you got to be thinking for it, too. You're thinking for the truck and the trailer.

"When you actually got used to it, it was kind of like playing dominoes or something. As long as you keep your mind geared for it, you're all right. But if you forget and quit thinking about the trailer, you could be in problems right now."

Little said drivers who had extensive experience on Lower 48 highways got in trouble when they assumed driving in Alaska was exactly the way it was on the Outside.

"You'd be surprised at the older drivers that have been around a long time, driving everywhere in the country, come up here and thought there's nothing to this. The next thing you know, they are in the ditch, upside down or maybe dead."

One of the winter hazards in the far north was driving when visibility was eliminated by blowing snow. This happened several times each winter, when drivers were forced to navigate by looking at stakes with amber and red reflectors along the side of the road.

"A lot of time you could not see from one stake to the other," said driver Fred Austin. "Sometimes I've literally driven by Braille, just creep along real slow and when you feel your front end drop then you back up. You can tell which side of the road you're falling off of. And you correct and you go on."

A bigger hazard than overconfident drivers and even winter whiteouts were the drivers without any experience who had somehow managed to get themselves dispatched by the union for a driving job.

"Usually a guy that never drove a truck before, he wouldn't

even get to Livengood the first trip and he'd have an accident, and he wouldn't be back because everybody's afraid of dying," Little said.

Fairbanks trucker Phil Menges remembered coming south one time and discovering an accident on the lower part of the Elliott Highway. The northbound truck had jack-knifed, though the damage wasn't too bad. It was the first trip the young driver had made in his brand new truck.

"He come down around the corner of the switchback and got a little bit excited I guess. It was a little slick but not really all that bad. He hit the brakes, got excited and jack-knifed and put the front of the truck over the road, over the edge of the bank," said Menges.

"The most tragic part of it was that he had just spent a lot of money to buy a truck, say $40,000 worth, and take it up there and before he even gets into the real bad road, he damages it," he said.

"Several times I had seen people on maybe their first or second trip and they'd have a problem, like spinning out and sliding backwards, which is one of the worst problems you can have, an awful, awful thrill. And they'll give it up right then. They'll say that's not for them," Menges said.

"There's a difference between people that can drive that type of road, and enjoy driving on that type of road, and just a regular highway driver," he said. "A lot of the drivers seem to like this money a little bit too much and get the idea that the longer they sit there the more they make for just sitting there."

Menges said he figured he belonged to another larger faction of drivers, one that liked the money but wanted to get the job done expeditiously. He thought that most of those who wanted to lounge around were gone from the scene by 1976.

Butch Rohweder, a trucker from Washington, said he came to Alaska just "to get that big paycheck."

"But, hell, this is great country," he told a reporter while his truck groaned during the climb up Atigun Pass. "Darned if I'm not tempted to stay. Sure, it gets a little rough at sixty below. Your brakes can freeze to the drums, or you can pop a drive shaft like a candy cane if you're not kind of tender with the gears. Make a wrong move on a hill, and suddenly you're driving an eighteen-wheel toboggan. That's how come some guys call this road the Kamikaze Trail. But it's the best trucking in the world if you know what you're doing."

Drivers from big cities and small towns all across the country shared that opinion.

"It's just like miners if they hear there's a gold strike, they all want to be there. And this has been the biggest strike for truck drivers in the world," said Little. "It was just like digging into a gold mine. It was there for a couple years."

"I got a home that's paid for now. I never thought I'd ever have that. That's part of the strike that I've managed to save," Little said.

Many of the Teamster line haul drivers earned $7,000 to $10,000 a month. Those who bought their own trucks, which were then priced at about $50,000, could lease them back to the trucking companies and pay for the trucks in five or six months. In the Lower 48, it usually took five or six years to pay off a truck, driving up to 175,000 miles. Trucker Tom Reed was among those who bought two trucks, with $10,000 down on each, and paid them off in five months.

He said there was no place else in the world where a driver could pay off a truck so fast. He said he had learned more about truck driving in two years on the pipeline than in eight previous years driving in the Lower 48.

Reed was looking ahead, beyond the boom that was gripping Fairbanks. "It's got too much of a boomtown atmosphere now, overcharging and everything high priced. I think in another five years this will be a fantastic place to live. I think once this whole boom is over there will be enough good improvements that are going to stay that it will really be helpful to this town," he said.

Driver Gil Forey, who had come from Montana, paid for his new truck in a little less than six months. He agreed with Reed about the prospects for Fairbanks. "This is all going to clear up after a while. It ain't going to be this rushed stuff they have here now. It's going to settle down. I think it's going to be good."

Forey said that most of the problems on the haul road were caused with "hot seat" driving, which was when a driver was assigned to a company truck and was not familiar with the brakes, steering, or how it handled. When he climbed into his White Freight Liner, he knew how it would brake and how it handled. That didn't happen with a "hot seat," he said.

Hot seat or not, driving a truck was lucrative because the Teamster contract that Carr had demanded and got from Alyeska included a provision for a guaranteed eighteen-hour day for those driving the Haul Road. This meant annual pay of $80,000 a year for some drivers working nine to ten months a year.

Carr had first negotiated this eighteen-hour provision in contracts when Teamsters were driving the North Slope on ice roads. The drivers often were stuck for days at a time, and conditions could be exceedingly difficult. Those were the days, as veteran driver Fred Austin put it, when drivers would dine on frozen peanut butter sandwiches and whatever else they carried in the truck.

The pipeline boom brought a change of fortune and a vast increase in the number of trucking companies, but as one official of Weaver Brothers pointed out, there was a catch that made the prosperity short-lived: "Sure the companies are grossing a fantastic amount, but they're also paying out a fantastic amount."

When the bubble of pipeline prosperity broke, many major trucking companies went broke or pulled out of Alaska. And Teamsters Union Local 959, which had been regarded as a monolithic force during pipeline construction, ultimately had to seek protection under Chapter 11 of federal bankruptcy laws because it overspent during the pipeline boom, and there was nothing to sustain it afterward. Non-union trucking firms emerged and the Teamster leaders who followed after Carr died in 1985 no longer had the "hammer" to shut things down.

The truck drivers who had seen what life was like before the pipeline had some sense that it was an unreal experience, never to be repeated.

"We really worked hard during that time, but I was old enough to know that the majority of time in your life is pretty much skim milk," said Austin. "The pipeline was like cream that wouldn't last. I felt sorry for some of the young folks because they thought that was how life was always going to be."

Neal Menschel/Alaska State Library
Sarah Jensen worked in the Sheep Creek camp as an oiler.

"The amount of energy that's directed toward a woman here is enormous. It must be what it's like for an extremely beautiful woman all her life."

— **Suzy Weschenfelder, recreation director at Glennallen pipeline camp**

WOMEN AT WORK

When Melva Miller arrived at Pump Station No. 1 on the Trans-Alaska Pipeline, the culture shock she experienced had little to do with her being an Athabascan from Fairbanks.

"I think I was the third woman in that camp," she said. "There were about two hundred and fifty guys."

The women who worked in the pipeline camps, who ranged in age from nineteen to over sixty, could never be anonymous faces in the crowd.

"It strikes home the first time a woman goes to eat," one female worker said. "She walks down a center aisle, followed by five hundred pairs of male eyes that all seem to reflect the same thought. This starts at 6:00 a.m. and continues through all ten hours of the day."

Melva was a "bullcook" at Pump 1, a job title that had originated with those who cared for oxen in logging camps.

She made sixty beds and cleaned rooms at the pump station construction camp.

A high school graduate who had attended one semester of the University of Alaska, Melva had joined the pipeline work force when she was twenty-two.

"I thought, 'This is okay until I make my big money, and then I'll quit,'" she said.

It didn't quite work out that way. While doing her daily chores in the trailer-like camp, Miller met some ironworkers. "It all started as a joke," she said. "A couple of guys in the shop suggested that I come down one night and they'd teach me to weld. I immediately loved it."

After putting in a ten-hour shift cleaning rooms in the camp, she went to the shop and began to learn about "working iron."

"They'd set up plates like I was going to take the test. I practiced and learned how to weld in the shack after work," she said.

Eventually she applied to become an apprentice in the Ironworkers Union, which put an end to her seventeen-week career as a bullcook.

Her story was told in newspapers across the country. Like many of the other women who worked on the pipeline, she was breaking new ground, taking on a blue-collar job at a time when there was no longer a consensus that a woman's place was in the home. In 1975, *Time* magazine named ten women as the "Man of the Year," and Americans debated the merits of the Equal Rights Amendment and "women's lib."

Women held about 10 percent of pipeline jobs in late 1975, with about half working in nontraditional blue-collar jobs. The female workers, both those in office jobs and those in the field, signed on to the biggest private construction job in history for the same reasons as the men. Money and adventure.

Libbi Bonnee, a divorcee with a five-year-old daughter, worked at Prudhoe Bay to support her daughter Avery. She had a week-on, week-off schedule, which she said was a great boon.

"When you're working a nine-to-five, you end up taking a sleeping child to the baby-sitter and picking a sleeping child up. Here I have every other week off and since I've been here, I've had more time with Avery than I had in the past two years."

Holly Reckord/Fairbanks Daily News-Miner
Grace Bellanti, of Boston, joined the Laborers Union
in Fairbanks. On the job near Pump Station 3, she
shoveled away gravel brought up by the giant chain
links of the roc-saw.

Most of the women were dispatched to pipeline jobs by
the Teamster, Culinary, or Laborers unions, but some, like
Melva, pursued other trades.

Her work on the pipeline started her on a thirteen-year career
as an ironworker that took her to job sites across the country.

"I love welding because it's new every day. There is
always something to learn. Welding is not difficult, but to do
it well requires work and constant improvement. And I want
to be good," she said in 1975.

She said at first she faced a lot of skepticism from other
workers and she had to prove herself every time she took a job.
Now married and the mother of two in Fairbanks, she looks
back on those years with pride.

"What's amazing to me is that everyone worked together.
We were all working toward this one goal, to get the pipeline
built. Seeing it done gave you a feeling of satisfaction," she said.

That they could derive a sense of satisfaction from a blue-
collar job came as welcome news to many women who joined
the pipeline work force. Even the humdrum jobs like making
beds took on some glamour because of the high pay. Enough
glamour to merit an article in *Cosmopolitan* on women pipeliner

workers headlined, "Daring Girls on the Alaska Pipeline."

"For women, the pipeline has become an eight-hundred-mile freedom trail, a hard-hat fairy godmother whose magic wand zaps dreams into reality," *Cosmopolitan* said. "The divorcee mops a hallway and, three thousand miles away, her children's college education is assured. An immigrant bullcook cleans a bathroom, moving steadily closer to the small Paris hotel she and her husband hope to buy. The dark-eyed girl driving a crew bus is one month shy of becoming a landlord with an apartment house of her own."

The arrival of *Cosmo* girls and other women on the job site was not without problems. There was considerable opposition from supervisors and fellow workers who thought construction sites were no place for women.

In the era when Helen Reddy sang *I Am Woman, Hear Me Roar*, friction took many forms. Women complained about discrimination, harassment, and sexual propositions. Men complained that some women weren't tough enough or qualified to handle the work. Such stories as the one about the woman who was dispatched to drive a dump truck, when her only experience was driving a taxi, gained wide circulation.

The cases in which women received special treatment from their supervisors were cause for resentment among men who otherwise might have been more willing to accept an integrated work force.

Steve Porten, who had run a crane at Coldfoot, said in a 1976 interview that there was competition among the men to attract the interest of the women and the supervisors showed favoritism to the women.

"There was one gal out there that had never worked construction before and she was made labor foreman. By her being made a labor foreman, several guys who had spent years and years and years in the construction industry as laborers were overlooked," he said.

"I suppose I have a biased opinion, but she was on the ins with the superintendent on the job," he said.

A man hanging out at the Chandalar Camp recreation hall was asked if the women worked hard. "Sure they work," he said with a laugh, "as much as anybody else."

"Nobody works much, do they?" the interviewer continued.
"Oh, there are times," he said.

One contracting supervisor offered this observation: "There are a lot of women up there that are absolutely incompetent in their jobs. But there are also a lot of men who are absolutely incompetent in their jobs."

When the pipeline was proposed in 1969, it was expected that the customary camp rules would apply: no booze, no drugs, and no women.

The first two rules were widely violated during construction, while the third prohibition fell victim to the social changes of the 1970s.

Requirements that women be hired for pipeline jobs grew out of the affirmative action plan mandated in the Trans-Alaska Pipeline Authorization Act.

Early on during the pipeline, the oil companies would send out news releases about the first, second, or third woman hired for a particular job. Despite the public relations effort, resistance took many forms. Sometimes women complained that the men on the job just wouldn't let them do the work.

"They want you to be their little pet chicken," said a dynamiter in the Laborers Union. Many people were not quite used to the idea that women could be dynamiters, truck drivers, or laborers.

"You have to fight, and I mean literally sometimes, to pick up anything heavier than a piece of paper," said one woman Teamster.

Yet, the times were changing.

"We never pretended we were going to correct the social ills of the world, but, by gosh—we have made an impressive start, and it's something that I am certainly not ashamed of," Senior Project Manager Frank Moolin, Jr., said of the pipeline record of hiring women. "There are women tractor operators, truck drivers, warehouse persons, flag persons, laborers, secretaries, clerks, a few executives, medics, and 'unisex' bullcooks. I challenge any other endeavor in the United States to show the strides we have made in employing women in positions they can handle, and there aren't many positions women can't handle."

Neal Menschel/Alaska State Library

Maggie Joyner was twenty years old when she took a pipeline job as a "bus swamper" in Valdez, earning $650 a week in September 1976.

The women who fought to get these pipeline jobs said that despite the affirmative-action plans, it was often difficult to get hired or to be taken seriously.

"Increasing pressure is being received from women's organizations and women individually regarding employment opportunities," Alyeska personnel manager Glenn Lundell wrote in a memo early in the project. "One of the major difficulties is that it seems our contractors in the past have been telling women applicants they cannot hire them because Alyeska will not allow women in the camps."

He said a memo was needed to "remove any question of Alyeska incurring any liability if any of these complaints are pursued formally."

At Lundell's suggestion, the company distributed a memo that said: "Alyeska does not have a policy of excluding women from field site jobs. Our camps were designed on the assumption that there will be employment of women in line camps. If women are dispatched for work or assigned in clerical or administrative positions, steps must be taken to accommodate them."

When construction peaked, about 2,800 women were working on the pipeline, out of a work force of 24,500. This huge imbalance added bizarre dimensions to the "battle of the sexes" as it was waged in the pipeline camps, where people were cooped up for nine weeks at a time or longer.

"The pressure is very severe," said Suzy Weschenfelder, twenty-seven, a recreation director at Glennallen. "It changes you. You know that everything you do is going to be the subject of a lot of conversation among the guys. And they've got the time to talk about it.

"This may sound strange, but the men are very courtly with you. But come to think of it, there was one slimy little creep who got out of line. I was walking down the hall once in Tonsina and he offered me $100."

She complained to the union steward and the "slimy little creep" was fired.

Being outnumbered in camps by nine or ten to one, women said they had to watch carefully what they said and how they dealt with men.

"For some women I know it's an ego trip for them to be around this many men and most of the men think it's an ego trip for women to be around so many men," said Pat Campbell, a laborer at Chandalar. "I don't happen to think so."

Many women mentioned that since they were in the spotlight at all times, any mistake was cited as proof that women didn't know what they were doing.

"When you're doing something that's new, a lot of men think it's cute. You're not taken seriously. Things are very sexual here. You don't even know somebody, and they'll just come off to you very sexually right away. It gets to the place where I don't even want to smile at a man because he thinks I want to go to bed with him," a female laborer said.

There were lots of men looking for female companionship. Some of them were just lonely and wanted to talk. Others wanted sex.

The sexual tension was a product of the law of supply and demand. In the social climate of the 1970s, a decade one writer described as the "high-water mark of the one-night stand," some people welcomed flings on the pipeline. That was, after all, the period in which *The Joy of Sex: A Gourmet Guide to Love Making* became a bestseller.

"There are hordes hungry to talk to a lady," said Weschenfelder. "There's no way of dealing with all of them. You could drown in it."

At Coldfoot, fifty-five miles north of the Arctic Circle, nineteen-year-old Lisa Wilson of Oregon was one of 55 women in a camp with 650 men.

"The only thing I don't like is that there are so few girls. The men treat you like some high queen. They tease you," said Wilson, who was making seven times as much money as she had before the pipeline.

A security guard at Pump Station 1 said that many of the men started out with "You're the prettiest thing I've seen all day."

"After hearing that a hundred times or so a day, you get pretty sick of it," she said.

On her first day in camp, she came out of her locked room and went to the dining room with other security guards and found it "frightening and overwhelming" to be "checked out by three hundred guys."

"I looked straight ahead and kept walking and mumbling to the lieutenant who was walking with me about 'Gee whiz, I wish I was someplace else,'" she said.

The only problems she had were with the Oklahoma pipeliners, who knew as much about obscene remarks as they did about welding pipe.

Teamster Barbara Tallman of Anchorage said that driving a bus of Oklahoma welders back and forth to their job site was the easiest job she ever had, but she had to put up with a lot for $1,200 a week.

"The driving is no problem. I earn my pay on the way in and out," she told a newspaper interviewer. "The crap that I take is incredible. The men are getting nasty now because we're near R&R. When they get back from R&R, things are real mellow for a while."

A college graduate with a degree in anthropology, she spent the rest of her ten-hour work day hooking a rug in the bus or going for walks.

In 1975, when women's liberation was just getting on its feet, it was unusual to see a female face in the Laborers Union Hall in Fairbanks.

She said the men assumed she would "find a boyfriend from the 798ers, but I haven't yet, and that really bothers them. It's just accepted that when a woman comes up she gets an 'old man' to avoid continual harassment if for no other reason."

Some women saw the entire pipeline through the prism of the women's movement, finding harassment, discrimination, and opportunity on the job. They learned, after several weeks that left every muscle sore, that they could handle tough jobs.

"The pipeline is a unique experience for women," said Barbara Maier, who worked as a laborer on a rock crusher in Valdez. "It is an avenue in which the revolution can speak. It is causing drastic change in economic conditions. For once, pay scales are basically equal."

Jud Roche of Seattle, a former school teacher who went to work as a laborer, told a Seattle newspaper that the pipeline was vital to the feminist movement because so many women were learning that they liked construction.

"Those buses are something else though. They all had a

stack of *Hustler* and *Playboy* magazines and the ceilings of some of them were literally covered with girlie pictures," Roche said. "They were really gross, especially the ones from *Hustler*, so I'd take 'em down without saying anything about it. Most of the men would sort of be embarrassed."

Roche also found the graffiti on the pipe, which included many "gross drawings," to be "worse than any I've ever seen in a toilet." Roche tried to steer conversation on the buses to matters of philosophy and the like, but the men always talked about cars, trucks, airplanes, and whatever else they wanted to buy with their money. There was one exception, however.

"At one point I and one of the men were shouting at each other over the Equal Rights Amendment," she said.

Roche said she valued the company of other women, but there weren't enough to go around. She and another woman worked with a crew of twenty men.

"Most of the women I encountered were so taken with all the male attention they were getting that they didn't want to spend time with another woman. That was an easy trip to get into and some of the women who came up full of the sexual revolution just couldn't handle it where it got to be a different man every night," she said.

There was a lot of talk about promiscuous behavior in the camps and the circumstances under which it happened. On a visit to Dietrich Camp, Geraldo Rivera, then a reporter for *Good Morning America*, said he was told a story from one worker that was "something like the old Hollywood casting couches."

"You find a lot of the foremen with a put-out or get-out attitude, you know," said Judy, whose husband was back in Anchorage.

Geraldo said he stayed an extra day at Dietrich because he was snowed in, which allowed him to attend a big party where "we sat around drinking officially forbidden wine and swapping stories on the trails that led us all to Alaska."

David Hartman, the host of *Good Morning America*, was impressed by Geraldo's tales of women on the pipeline. "Those gals are teee-riffic," he said.

Alyeska officials also liked to portray the presence of women as a "teee-riffic" influence.

"Everyone seems to think that having women around makes life up here more civilized," said Alyeska spokeswoman Beverly Ward. "We noticed a change in the men immediately— that they were shaving more regularly and cleaning up every day."

Learning the nuances of construction culture took some time, however. Kelley Hegarty, a Berkeley graduate working as a laborer, was assigned to a crew mixing concrete. She met a Laborers Union official named Jim Sampson, who told her that a "safety meeting" had been called for that afternoon.

Sampson, a future state labor commissioner and mayor of the Fairbanks borough, told her about the idea of a "safety meeting," which was a way of stopping work without striking. He told her everyone would stop working until the contractor cleared up a problem on the job. But Hegarty told him she didn't want to go to the meeting. "I came up here to help build this pipeline," she said.

"Oh, you did?" said Sampson. When it was time for the safety meeting he returned with two of the burliest men she had ever seen, who politely escorted her to the safety meeting.

An in-house Alyeska report about women on the pipeline said that as the project went on, "managers' views have changed from anti women in camps to very definitely pro."

Alyeska Section Manager Dave Haugen said he would not want to go back to the old way of male-only camps. "Part of the success is due to the high caliber of women. The standard of behavior is much higher than in a small town. Craft people, especially women, are young. There tends to be quite an age difference between men and women. The older men are protective towards the younger women, many of whom are the same ages as their daughters."

Not all of the older men had this attitude, however. Some of them, and the younger ones as well, figured that women who went to work on the pipeline were looking for men.

"My wife said she wanted to work in a camp to see what it's like," a fifty-year-old construction worker told an interviewer. "I told her, go ahead if that's what you want. But don't expect me to stick around. I've been there, and I know what those broads want."

While he may have claimed to know "what those broads want," many women took great offense at the suggestion that they were on the scene to provide entertainment for the troops.

"I am here to work, to make money, like everyone else," said a woman at Franklin Bluffs. "I am a shy person, and it is very hard for me to adjust to camp life. I have made friends with the men on my crew, and they are the ones I relate to. I try to be friendly to everyone, but it is a strain for me to relate to several hundred men. And I'm used to having more women to talk to also!

"It is not, as one may think, absolute heaven living and working in an environment when the ratio of women to men is so small," she added. "It is unfair to classify and judge us all. I don't feel so special. I am insecure and homesick and doing the best I can handling this new and different lifestyle."

Angie Reilly, a radio operator at Coldfoot, said she thought the pipeline was a once-in-a-lifetime opportunity, but the high paychecks came at a price.

"Because I love my family, I'm willing to make sacrifices. And it is a sacrifice to be away from them. I missed the first perfect bow my daughter tied in her shoes," said Reilly, who had school papers and family pictures all over her room.

"I'm not doing it to show I can go out and do all the work a man can. I'm not a women's libber in that sense. But I think a man is just as important at home as a woman, so there's not that much difference between a woman leaving her home and a man leaving."

"Alyeska wants a pipeline. To lay one, you've got to have my kind of people. It takes a pipeliner to do this kind of work. You can't take a carpenter out here and have him lay pipe. Or a bricklayer. Or a dirt hand. You've got to have a pipeliner to lay it. I've been on a lot of jobs. This is all I've ever done. I've never done anything else but pipelining. It's my life and my whole family. And all of my relatives too."

— **An Oklahoma welder from Local 798**

THE WELDERS FROM TULSA

The members of Pipeliners Local 798 from Tulsa, Oklahoma, became famous in Alaska for welding pipe and raising hell. They excelled at both.

Whenever there were fights over the quality of the cuisine, the availability of washing machines, or about which union was the most indispensable, likely as not, 798ers were in on it.

Alyeska officials said that the pipeliners, who came from many states in addition to Oklahoma, considered themselves the "Marine Corps" of the labor movement. They specialized in pipeline work and were a tough bunch, regarded by many as prima donnas.

"We've got a few Yankees among us, but not many," said Wayne Throckmorton of Arkansas.

One often-repeated rite of initiation was to "grease" a newcomer, which meant dropping his pants and giving him a shot of grease in the genitals.

"That was just a big joke," said Dwayne Williams, a 798er from Oklahoma. "That's kind of a thing of the past now. It was a nomad-type work," he said, and at that time many of the 798ers tended to be on the rowdy side.

"We're no different from other people," said welder Dick Gibson, a worker at Tonsina, who was angry about how Local 798 was portrayed in the media. "We go to church and belong to the same organizations everyone else does. This is just another pipeline for us. There's not another local that can supply the men with the experience necessary to do the job."

Alyeska President Ed Patton agreed: "You can teach any number of men or women how to operate a D-9 bulldozer satisfactorily within a reasonably short period of time, but I know of no way to teach the skills of pipeline welding within the time span of a normal project."

The 798ers wore bright-patterned welders' hats, jeans, and cowboy boots. They wasted no time establishing a reputation that led to the "Happiness is 10,000 Okies going south with a Texan under each arm" bumper sticker. (The simple follow-up to that one was another bumper sticker that said, "With $20,000 in each pocket.") Many of the pipeline workers from the South agreed with that idea of happiness.

"It's great to be from Oklahoma, but much greater to be back when it's over," said John Marta, a former dinner club operator turned pipeline inspector. "There is nothing like catfishing, poke pickin', wild onions and blackberries, with a few chiggers and ticks thrown in for excitement," he said.

The anti-Okie sentiment was not just a bumper sticker to Linda Rowan, a 23-year-old Oklahoman who had been in Fairbanks with her 798er husband for two years.

"There are those few who detest us pipeliners," she said, describing the welcome she had received in Alaska. "Most aren't associated with the pipeline at all, and I can understand their feelings to a certain degree. But there is no call for being made fun of for having a southern accent, being called a 'redneck,' or having obscene gestures made at you while just driving down the street."

The 798ers made $18.25 an hour straight time, far above the hourly rates for the other unions. "I thought about being a brain surgeon, but it wouldn't pay me enough," said Keith Duncan, a twenty-five-year-old pipeliner from Bald Knob, Arkansas.

It wasn't unusual to have fathers, sons, brothers, cousins, uncles, and in-laws on the same pipeline welding crew. Those who weren't related knew their fellow 798ers from other jobs, for the local's members had welded most of the gas and oil pipelines in North America.

A welder and helper at Dietrich in August 1975.

"They tend to keep to themselves, cultivating a reputation for being hardworking, hard-drinking, skillful guys who don't have to answer to anybody," wrote pipefitter Potter Wickware.

Wickware, a member of Local 372 of the pipefitters' union in California, was among that large assemblage of welders from other union locals who handled the equally important and complicated welding at the twelve pump stations and the Valdez terminal. The 798ers handled only the four-foot pipe.

"Take your typical 798 pipeline welder and feed him a few drinks, and he'll probably tell you that he's God's greatest gift to welding," Wickware said.

Or as one of their members from Arkansas put it, "I can weld anything from a broken heart to the crack of dawn."

Peaceful resolution of labor disputes was something else,

however. To settle one disturbance north of Fairbanks at the Livengood pipeline camp, three planeloads of Troopers and security guards arrived in riot gear to arrest a Texas pipeliner. He had raised a ruckus after showing up a half-hour after the kitchen had closed and demanding a hot meal.

Trooper Jay Yakopatz remembered another instance when a security guard who asked a 798er for identification was knocked to the ground in Fairbanks for his trouble.

"The pipeliner stomped all over him, then went into the mess hall for a gang of his buddies to help him take on the security officers," said Yakopatz, who was a shift corporal for the Troopers. "The Military Police asked for Trooper backup and we rolled every man that was available. The situation was finally brought under control, and the troublemaker arrested. Later, some of his friends tried to break him out of the state jail."

Dana Stabenow, who later became one of the state's most prominent mystery writers, worked as an innkeeper for Alyeska at Galbraith Lake, north of the Brooks Range. The first lesson she learned was to never put a 798er in the same trailer as a Teamster. On one occasion, a welding crew filled up one fifty-six-man trailer and there were two additional welders assigned to another trailer that wasn't very full. The welders proceeded to kick everyone who was not a welder out of that section of the camp.

In another warning, Alyeska auditors had been critical of a welding crew's performance and alleged the welders were trying to get paid for hours they hadn't worked, leaving the job a half-hour early and getting the bus to make frequent stops so they would get paid for more travel time.

Tensions escalated when the welders shifted into low gear, an action known in the trade as a "slow wobble." When Alyeska officials did not relent, about seventy welders quit en masse. The welders gathered their gear, commandeered several buses, and began heading toward Fairbanks. A union steward learned what had happened and caught up with the workers on the Richardson Highway and persuaded the welders to return to camp. But Dave Haugen, an Alyeska supervisor on that section of the pipeline, refused to allow them back in the camp.

Al Williams, a 798er from Ocean Springs, Mississippi, welds pipe in the Thompson Pass section, just north of Valdez.

"They had quit, and as a result we had no obligation to negotiate with their union leaders over any grievances they might have, nor did we have any obligation to rehire them," Haugen said.

He ordered the guards at Isabel Pass Camp to lock the gates and send the buses north on the Richardson Highway to Delta, where the crews would get their final checks. In the meantime, Alyeska called the Alaska State Troopers and warned that the confrontation in Delta would be "emotionally charged and possibly violent."

The buses arrived at 6:00 p.m., and the men stayed in them outside the locked gates of Delta camp until 9:00 p.m., when they received their checks and headed to Fairbanks. One man was arrested in a fight as the buses prepared to leave.

That same day there was a drunken brawl at Fort Wainwright in Fairbanks that led to the arrest and eventual conviction of six 798ers. Five guards, two MPs, and an Alyeska official were injured in that brawl, which started when a security guard was assaulted by a pipeliner.

Battles of this kind became standard newspaper fare during the pipeline. The most dramatic fight took place north of Valdez at the Tonsina camp.

The tussle in Tonsina triggered a showdown between the 798ers and the Teamsters Union and a temporary work stoppage by four thousand workers on the south part of the pipeline.

Robert Williams, a Teamster bus driver, was driving a busload of welders back to the pipeline camp after a night of carousing at the Tonsina Lodge when the trouble began.

"Everybody was real rowdy," he said. "Two or three times I almost had to pull the bus over because of fights."

One of the fights he saw in his rear-view mirror was between two or three welders' helpers and a laborer. The welders were boasting about how the pipeline couldn't be built without them, while the laborer was asserting that it couldn't be built unless all the trades worked together.

Reaching the camp, some of the men were still arguing with the laborer, and Williams grabbed one of the welders by the arm, because he thought the man was going to hit the laborer. The welder's shirt was ripped when Williams grabbed him.

A couple of days later, Williams was at the pipeline camp when he heard an announcement over the PA system asking "all 798 hands to report to the shuttle bus," which Williams was supposed to drive. A mob of about a hundred men encircled him, and the union steward said Williams should apologize for ripping the shirt and pay the welder $100, or he would be "worked over."

Williams refused to pay, and four or five guys beat him up. Members of the Teamsters Union walked off the job for two days in protest of safety conditions. The union steward later pleaded no contest to an extortion charge.

Union officials said there was less fighting, drinking, and gambling on the pipeline than they had expected, but that the fights were more noticeable because the workers were concentrated in the camps. "Physically we're no tougher than anyone else. We have a group of people that know how to take care of their selves," said Union leader Bill Lancaster.

L.E. Burnett, a welder from Marble Falls, Texas, said that fighting was a way of life for many young men he grew up with. "I've been in fights all over this country and I didn't win all of them," he said.

In addition to their pugilistic prowess, members of Local 798 also found themselves in the spotlight after a scandal broke about a contractor who had falsified some of the X-rays taken of the welds.

"We're getting it in the shorts," said Pee Wee Ingram, a welder's helper from Liddleville, Louisiana. "Alyeska's lied about every damn thing from the quality of the weld work, to the steak and lobster they're supposed to be feeding us."

The stipulations required that every pipeline weld be X-rayed, but the X-ray crews couldn't keep up with the welders during the early part of construction.

Someone decided to eliminate the X-ray gap by using falsified X-rays, which raised a host of expensive questions about welding that didn't get settled until just before the pipeline went into operation. The reinspection and repair of a relatively small number of welds cost from $80 million to $100 million, according to one estimate.

Bobby Seales, a 798er from Alabama, said when he went home on R&R about twenty-five people asked him, "Bobby, what's wrong with those pipeline welds?"

"Well, I told 'em there's a whole lot of misconception goin' on," said Seales, who spoke with a mouth full of Copenhagen.

"There's nothing wrong with the welds. The trouble's with all the paperwork and dumb inspectors Alyeska's got."

Two decades later, it's still not clear what was gained by the expensive weld repair program, and how much of the fiasco was about welding quality, or simply the interpretation of X-rays and record-keeping.

"In the real world, many of the dollars were spent to take care of something that had no significance," said Jack Turner, who was the chief federal surveillance officer at the end of the project. The head of the state surveillance agency said much the same thing.

The last of more than 100,000 pipeline welds was completed May 31, 1977, about a thousand feet north of Pump Station No. 3, performed by "Rat's Tie-In Crew" before a crowd of about two hundred workers and dignitaries.

After the pipeline was built, the nomadic welders left Alaska, bound for the next pipeline job.

SECTION II

SKINNY CITY

Alyeska Pipeline Service Company

Two-story dormitory housing at Pump Station No. 1 housed up to 450 North Slope workers.

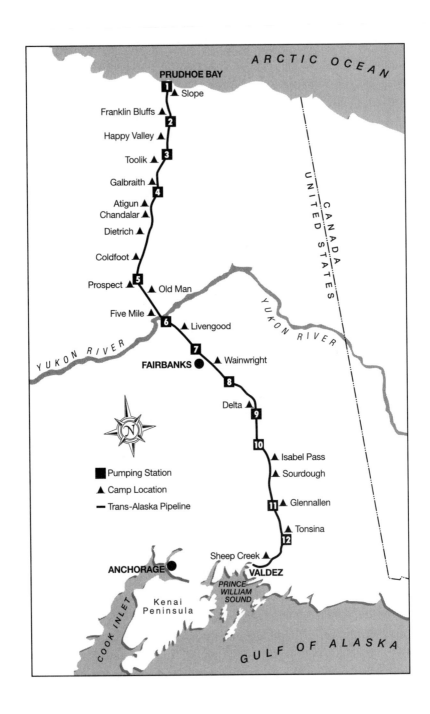

THE CAMPS

Spread out along a route equal to the distance between New York and Chicago, the pipeline camps ranged from the compact 250-person models at the pump stations to the mammoth 3,500-person terminal camp at Valdez. The typical pump station camp cost $6 million, while the typical mainline camps cost $10 million to build.

With the exception of Wainwright, which consisted of U.S. Army facilities in Fairbanks, the camps were all of standardized modular construction. The basic housing units had twenty-eight rooms, with space for fifty-six workers in each wing. The pump station camps and the one at Valdez were two-story affairs. The rest were one-story.

The pipeline starts at Prudhoe Bay, named in 1826 by Sir John Franklin in honor of naval officer Algernon Percy, the first Baron Prudhoe. It ends at Valdez, named in 1790 by a Spanish explorer in honor of naval officer Antonio Valdes y Basan.

The camps along the route included:

Slope
Originally named Surfcote, after the company that was contracted to put a protective coating on the pipeline, it was renamed in 1974.

Franklin Bluffs
Named by the U.S.Geological Survey for the bluffs on the east bank of the Sagavanirktok River in 1951, probably in honor of Sir John Franklin, a British captain who explored the north coast of Alaska in 1826.

Happy Valley
Said to have been named by a survey superintendent who said, "This looks like a happy valley to me."

Toolik
From the Eskimo name for the common loon and the yellow-billed loon and often spelled *tudlik* or *tulik*.

Galbraith
For Galbraith Lake, which was named after bush pilot Bart Galbraith, who was killed in a 1950 plane crash while flying from Barter Island to Barrow.

Atigun
Taken from an Eskimo name for the Atigun River in the Brooks Range.

Chandalar
Named for the Chandalar River, which came from the French *Gens de Large*, meaning nomadic people, a term French employees of the Hudson's Bay Company applied to aboriginal residents near Fort Yukon in the late 1800s.

Dietrich
Named after the Dietrich River, which may have been named after a miner. The name was listed on a field sheet in 1899.

Coldfoot
Named for a town that was first called Slate Creek before green stampeders in 1900 turned back after getting cold feet. In 1902, according to writer Bob Marshall, the town consisted of "one gambling hole, two roadhouses, two stores, and seven saloons."

Prospect
Named for nearby Prospect Creek, one of six streams in Alaska with that name. A temperature of eighty degrees below zero Fahrenheit was recorded here on January 23, 1971, the coldest on record in Alaska.

Old Man
Named for the Kanuti River, which was called *Old Man River* by some prospectors in 1898-1913.

Five Mile
Named because it was five miles north of the Yukon River.

Livengood
Named for Jay Livengood, a prospector who discovered gold nearby in 1914.

Wainwright
Located in facilities leased from the U.S. Army on Fort Wainwright near Fairbanks.

Delta
Named after the nearby town of Delta Junction, which was named after the nearby Delta River.

Isabel Pass
Named for Isabel Pass, which was named for the wife of E.T. Barnette, the founder of Fairbanks. It housed up to 1,652 workers, second only to the camp at Valdez.

Sourdough
Used for eight months, the shortest of any of the construction camps, it was built in the winter of 1975-76 because there was a lot of specialized work between Isabel Pass and Glennallen.

Glennallen
For the town of the same name, which was a combination of the names of early explorers Captain E.F. Glenn and Lieutenant H.T. Allen.

Tonsina
Named after the Tonsina River. A telegraph station was established at Tonsina in 1902 by the U.S. Army Signal Corps.

Sheep Creek
Named for a nearby stream, one of twenty-eight streams in Alaska called Sheep Creek.

The Pump Stations

The pump stations contain aircraft-type jet engines that power the turbines that drive the powerful pumps that force oil through the forty-eight-inch pipe at about five to six miles an hour. At peak production, ten of the twelve pump stations were in use. Pump Station No. 5, north of the Arctic Circle, does not have pumps. Its storage tanks are used to store oil when needed. Pumps also were not installed at Pump Station No. 11, one hundred and fifteen miles north of Valdez.

The biggest problem during the startup of the line occurred when Pump Station No. 8 outside of Fairbanks blew up, killing one worker. The station was bypassed while rebuilding began, and the first oil reached Valdez at 11:02 p.m. on July 28, 1977.

Pump Station	Pipeline Milepost	Elevation in feet
1	0	39
2	57.58	602
3	104.29	1,383
4	144.60	2,763
5	274.75	1,066
6	354.96	881
7	414.15	905
8	489.24	1,029
9	548.71	1,509
10	585.79	2,392
11	685.88	1,302
12	735.06	1,821

Atigun Pipeline Camp in the Brooks Range was north of Atigun Pass, at 4,739 feet the highest point along the pipeline route.

"When you feel that one day you're going to look down and fly across or whatever and there won't be anything here, it gives you an odd feeling, that the place you got to call home is going to be gone entirely."

— **Construction supervisor Dean "Ol Grizz" Elston, in a 1977 interview talking about the impending removal of the camps built for workers on the Trans-Alaska Pipeline.**

AN EIGHT-HUNDRED-MILE NEIGHBORHOOD

They are the ghost towns of the Trans-Alaska Pipeline.

Once they were twenty-nine miniature cities spread from Prudhoe Bay to Valdez, home to every virtue and vice. Or, as one reporter put it, they were collectively "Skinny City," which was eight hundred miles long and a few hundred feet wide.

The neighborhoods of Skinny City included Happy Valley, Franklin Bluffs, Galbraith, Dietrich, Old Man, Prospect, Isabel Pass, Sourdough, Five Mile, Slope, Fort Wainwright, Delta, Glennallen, Tonsina, Toolik, Chandalar, Sheep Creek, Livengood, and the twelve pump stations.

For some workers, living in Skinny City was Fat City, with room and board costs paid entirely by "Uncle Al," as Alyeska was known.

"Well it's good living, like I told you," said an old Alaska Native man working as a construction hand. "It's like Pioneer Home. You got good maid service, everything is clean. Clean life. I never lived like that all my life."

The camps had a total of 16,500 beds and were home to about 60,000 people over three years. Today, evidence of their existence is harder to find than some Gold Rush boom towns.

When the gold panned out, no one removed the remnants of the boom towns that became ghost towns. What was left fell victim to vandals, weather, and time. In the case of the pipeline camps, Alyeska was required by law to remove them.

All that remains today, with the exception of facilities like the Arctic Acres Inn Resort Motel at Coldfoot, which was a former pipeline camp, are gravel pads and some landing strips.

The last camp to open, and the first to close, was Sourdough, located between Isabel Pass and Glennallen on the Richardson Highway.

They weren't really cities in the normal sense of the word. They were temporary camps, with temporary residents. The population changed rapidly, with people quitting and going on R&R and getting dispatched to other camps constantly.

In a real city established routines help give substance to everyday life, trips to schools, to the grocery store, to church, and a place to call home. In the pipeline camps, the normal social structure was replaced by a hybrid that revolved around working, eating, and sleeping, and two-to-a-room cubicles in a trailer.

"Our rooms may not be as large, but they're as comfortable as anything you'll find in a good hotel," said Coldfoot superintendent Kent Gildesgrad.

From the air, the typical camp looked like a cross between a sparse mobile home park and a heavy equipment yard. The long, rectangular pre-fab housing units would be lined up like building blocks and usually connected by enclosed, heated walkways.

The camps, built on pilings or skids with above-ground water and sewer pipes, had a temporary look and feeling to them in keeping with their origins as transplants in the wilderness. Powerful generators hummed in the background, and the smell of diesel fuel exhaust was in the air. The lights burned through the long winter nights in kitchens, warehouses, offices, and garages.

It cost Alyeska anywhere from $50 a day to $200 a day to house a worker, according to various estimates given to author Robert Douglas Mead at the time.

Mead wrote that "like other Alyeska costs, it is elastic, varying with the speaker and his purpose."

There were washing machines for the workers to clean their clothes and "bullcooks" who provided maid service.

Mead wrote that a camp he visited looked like a "raw and

utilitarian college campus where all the courses are practical, the men and a few women mostly young and hairy, on their own, with defined and circumscribed purposes."

Some of the camps had enclosed hallways and some did not. It was a matter of debate as to which was better.

"I think the Arctic hallways give a lot of people sort of a claustrophobic type feeling. It's unhealthy in a way that a lot of people can live there for months and never go outdoors for anything. The food, recreation, their bed, their job, everything is right there," said six-foot-three Mike McDermid, the camp manager at Dietrich.

"The camp that I enjoy most is Chandalar, where they don't have any Arctic hallways. But they're perched on the side of a hill and have a fantastic view. They have probably more sunshine over the year than any other camp," he said. "People are required to go outdoors to go to eat, to go to their job and so on. I think that's a healthier, better situation."

The problems of cramped camp life were magnified in winter, especially in the northern camps, where nearly the entire day was spent in darkness and frigid temperatures.

"Right now with the cold and the darkness, everybody is looking for a reason to go home," he said one January day. "People are miserable. They get crowded. You get two guys in the same room. One complains it is too hot. The other says it is too cold. It is probably neither one. The truth is, they are restless."

McDermid said that some of the discipline problems reported at camps like Franklin Bluffs were due to the enclosed living quarters. But, he said, even if people went outside at Franklin Bluffs on the North Slope, they wouldn't have much to see or do.

In the Brooks Range it was different. He said that at Dietrich and Chandalar and other camps, some workers brought their cross-country skis to get exercise after work.

Not everyone shared his view that the camps without enclosed hallways were the best. A welder from Muskogee, Oklahoma, said the camps with the closed-in halls were superior. After work he liked to take a shower and have a few drinks in his room, like a lot of other workers who ignored the no-

Steve McCutcheon/Alaska Pipeline Service Company

Five Mile Camp was so named because it is five miles north of the Yukon River.

alcohol rule. The Okie from Muskogee said he didn't like putting on heavy clothes to have to go to the chow line or to pick up his mail.

For some workers, life in the camps helped them preserve their pipeline paychecks because the legal options on how and where to spend money were limited.

"All you're out is cigarettes, razor blades, such as that," said a Oklahoma welder at Galbraith Lake. "That's one big advantage."

The lack of check-cashing facilities was a disadvantage, he said. "You can't play too much poker because you can't get a check cashed up here," he said.

This wasn't a hindrance to anyone who was serious about playing cards, however. It was easy to find a poker game or more elaborate gambling taking place.

"When I was at Galbraith Lake camp last week, a full-fledged craps table with a professional-sounding croupier was in operation in one of the rec halls," Alyeska spokesman Larry Carpenter wrote in a 1974 memo. "There also is evidence of more than casual gambling in other camps."

Poker players at Delta camp anted up $100 to get into one six-hour game, with a minimum raise of $100. On this occasion, five men, all in their forties, sat at the table after a twelve-hour shift.

"You just watch the way we bet," one of the players said. "It's only paper—green paper money, just like my kids' Monopoly set. You make your money up here, and there's not much you can do with it. You can drink it. You can gamble it. You can whore it. And you can burn it."

A reporter from the *Washington Post* watched as the worker lit a $100 bill, then ground it out and put it in the pot as his ante.

One of the players, Larry Wills, a welder from Tulsa, was up $2,100 at one point in the game. He was down $1,700 by 2:00 a.m., which was $200 more than his gross pay the week before. For workers who gambled, payday was a highlight and many of those evenings ended in fights.

"You make more than you're used to, and you piss it away," he said. "Like the snow turns yellow for a minute, and then more snow covers it up. The green is there, and then it's gone. Some nights I win at cards, and some nights I lose. This week is bad. Maybe can't send my wife a check next week."

He thought he might tell her that he had been docked seven days pay for some misdeed on the job by way of explanation. "I gamble," he said. "But I don't run around with women. I'm not drinking myself to death like these guys who come up here and pour their money down their throats till they pass out or puke their guts out all over the hall. And I fly home to be with my family every nine weeks."

The gambling and drugs to be found in the pipeline camps occasionally prompted visits by Alaska State Troopers. Trooper Bob Stewart of Deadhorse once discovered a pool in which those who entered were guessing the number of feet of pipe to be laid in a day. The prize was a trip for two to anywhere in the world "with the assurance that the winner's job would be covered, and he would continue to receive wages while he was gone," reported one written history of the State Troopers. There was more than $100,000 in the pot when Stewart broke it up.

Trooper Wayne Schober, who flew a military surplus Beaver aircraft to the camps, said that news of his impending arrival with a drug-sniffing dog always traveled fast.

"Several times the dog located major caches of narcotics in the camp living quarters," he said. "One of the managers told me they received word the dog was enroute the moment we took off from Fairbanks, and the word flashed up and down the line of camps. There was, according to him, also a major loss of water pressure due to toilet flushing on these occasions."

Drug use in the pipeline camps was common, particularly among the youngest workers, because the drug culture was part and parcel of the "Me Decade" in America. This was especially true in Alaska, where the Legislature had decriminalized possession of small amounts of marijuana and the state Supreme Court had ruled that citizens had a constitutional right to possess marijuana in the home.

Smoke alarms in the rooms of some camps didn't react to cigarette smoke, but they were set off by clouds of marijuana smoke. Some workers solved this problem by putting wet washcloths over the alarm.

Precise figures on drug use are not available, and opinions vary about how widespread it was. A worker who had been employed in more than half of the camps at different times wrote about drug use on the pipeline under the pen name "George Piper."

He said that not everyone used drugs, but there was a "laissez faire" attitude about the whole thing. Crackdowns occurred when drug use became an obvious work problem. Cocaine was clearly illegal and available, but it was not as common as marijuana, which was probably second to alcohol as the drug of choice.

"Given a safe opportunity," Piper wrote, "as much as 25 to 30 percent of the work force might smoke during work hours. A truck cab, the dump, a hidden side of a building—and the always available john—provide havens for a quick smoke. But unless a worker knows for sure he won't be spotted by other than friendly eyes, the risk of getting fired from that $500 to $1,000 per week take-home-pay job is just too high for many."

He quoted a laborer foreman at Happy Valley who said

the most likely people to smoke marijuana were laborers, some oilers, Teamsters, kitchen workers, and bullcooks.

"Either the endless repetition, like shoveling; or waiting time, like a bus driver having nothing to do between when he drives workers to the job site and when he goes back to camp— those are the type jobs where people find it easy—even necessary to smoke," the laborer foreman said.

Many of the older construction workers wanted nothing to do with the younger pot-smoking crowd. Tom, an old construction hand at Pump Station No. 1, told about his new young roommate. "First thing he asked was, did I smoke. I said no, but go ahead, it don't bother me. It turned out what he meant was pot. Well, I told him. Next day he moved someplace else."

Still, other workers thought of marijuana and cocaine the same way they did of alcohol. A common sentiment was expressed in a poem by Don Crisman, who wrote from Franklin Bluffs:

> *The camps are all dry, or so they say.*
> *But I've been here forever, a month and a day.*
> *And we've all got something, be it booze or dope.*
> *Cause time goes slow for the men on the Slope.*

For almost everyone except for those who tried to go as long as they could without taking a break, the date that your next R&R began was as easy to remember as your birthday. Wall calendars in many a room were marked off day by day.

Wien Air Alaska, which flew Fairchild F27s and FH227 aircraft, was the only carrier with scheduled service to the camps north of the Yukon. The sound of the "Wienie Bird" was the sound of freedom for many a pipeliner far from home.

A NANA Corporation security guard at a pump station said that the men who stayed in camp for three or four months without leaving were going "crazy" by the time they left, but even those who stayed only the recommended nine weeks felt the strain.

"By the time the nine weeks were over, they were walking around like zombies," said Chaplain Ray Dexter. "You could tell stress levels were building up as time went on."

Ken Roberts/Alaska State Library

Former gold miner Ross Brockman, of Wiseman, plays Rudy Vallee on his Gramophone in October 1975. Like others from another boom in Alaska's history, Brockman's life was fairly quiet until the pipeline came along. Dietrich camp was built about twenty miles north of Wiseman, where the winter population was about four.

Part of the stress came from putting people together who in normal life wouldn't necessarily be living in such close quarters. There were college graduates with advanced degrees sharing space with people who had dropped out of high school. They ranged in age from teenagers to men and women in their fifties.

Dwayne Williams, a welder from Cleveland, Oklahoma, said he had welded pipe in all but three of the fifty states, and Alaska was unique because nowhere else were you so isolated. "Down here you stay in a motel room," he said from his home in Cleveland. "It was great because it was all free. Unless you wanted to gamble, you could send all your money home. A small percentage gambled it all away. As a rule I think it was a good moneymaker for everybody."

There was evidence of pride by some of those who thought of the camps as a temporary home.

"It's a lot easier to live in this camp than it is in Fairbanks," said Paul Frerichs, a timekeeper. "I took my R&R in Fairbanks and I was glad to get back to Coldfoot."

A Caterpillar operator from Washington, who was working at Coldfoot, said his main complaint was that the unions required him to take R&R after eighteen weeks.

"After all, we're all up here for the money," said Jim Robertson. "The life here is OK, and after you go out, it's tough to come back."

Fred Stickman, Sr., sixty-seven, a man from Nulato who was famous throughout Alaska for his pointed insights in letters to the editor, liked his R&R, but he also liked the camps.

"I don't like to go home as I have to cook for myself, wash dishes, wash clothes, pack water, haul wood. About half the time you cannot buy the kind of grub you want. So I'd like to spend the winter somewhere here on the Slope," he wrote once from Old Man.

Another worker with a long view of life in remote camps was Dorothy McGonigle, the first woman working at Prudhoe Bay. A widow with two grown daughters, McGonigle said her job handling field records was a pleasure compared to the fifteen months she had spent at a camp in the Northwest Territories during World War II. The earlier job had "no telephones, no telexes, only Armed Forces radio, no fresh fruits and vegetables, no inside plumbing," she said.

Those who had the inclination to add other activities to routine camp life had the opportunity to pursue a wide variety of hobbies in their spare time. Randy Putz, an Alyeska contracts engineer at Galbraith, spent an average of three hours a day building a model ship, finishing the job in about three months. There were many impromptu music groups, mainly playing country and folk music. There were classes in karate at Delta, instruction in soapstone and wood carving at Dietrich. At Isabel Pass, timekeeper Shirley Hughes directed plays in the "Pipeline Playhouse," starting with a rendition of *The Sunshine Boys*.

The camps were made of nearly identical pieces put together in similar ways, but there were differences among them. The theater at Chandalar had many pictures of actress Marlene

Dietrich, thanks to fan Rudi Polot, a naturalized citizen from Austria who was the camp's head bullcook.

Coldfoot had door pulls and coat racks in its dining room in the style of the old mining camps. Chandalar had tablecloths of red, blue, and yellow, and a reputation for being "mellow," while Franklin Bluffs had a rowdy reputation.

Atigun, the smallest of the original pipeline camps, had a more "family" type of atmosphere, said worker Steve Fromme. It had a population of about 300 and was known by some as "Atigun High School."

The road sign that marked the "Atigun City Limits" identified the Brooks Range camp as the "Shangri-La of the Trans-Alaska Pipeline."

"Of the many things Atigun was, it was not a sleepy little camp in the Brooks," he wrote just after the camp closed for good. "As many former residents can testify, something always seemed to be going on. It might be a game of volleyball at Chandalar, a softball game at Galbraith or perhaps a dance in the theater. (Oh! Remember those dances!)"

The volleyball team at Atigun was known as the "Slopetrotters," led by Bob "Bimbo" Nalow, the camp's "resident garbologist." The Slopetrotters claimed to be the best in the northern section of the pipeline. They wore bicentennial uniforms of tie-dyed T-shirts in red, white, and blue.

There was also a sign to the "City of Five Mile," and a sign beneath a tree that stood three and one-half miles south of Chandalar. It said, "DO NOT CUT THIS TREE. FARTHEST NORTH WHITE SPRUCE ON PIPELINE CORRIDOR."

There was an official sign welcoming people to Prospect Creek Camp, sixteen miles south of the Arctic Circle, and an unofficial one marking the "City Limit" of Prospect Creek. The notice identified Prospect as "Home of the Meanest, Workingest, Sons of Bitches in the Far North. Flimflammers, Rapists, No-Good Drunks and Red-Necked Pecker Woods 'Welcome!' However, absolutely no admittance to blademen from Old Man."

On the subject of signs, Joe Milligan, a driver for Green Associated at Old Man Camp, painted a four-by-eight-foot sign that said "Welcome to the Arctic Circle." It eventually was

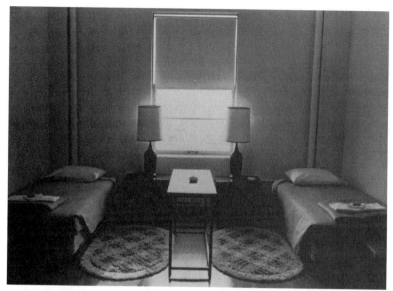

Charlie Backus/Alaska State Library

Five months before the first piece of pipe was laid, Alyeska Headquarters, then at Fort Wainwright near Fairbanks, displayed this model of a pipeline camp room.

covered with dozens of signatures from those who stopped along the haul road at the sign.

Construction hand Allen Chesterfield, who had eighteen pipeline jobs in 1975-76, said the first time he saw the Alyeska camp at Delta, he realized that each ten-by-twelve-foot room featured furnishings the company took to be the "comforts of home." This included red-and-orange carpets, bright orange curtains and bedspreads, hard wooden chairs, a small desk, mirror, extra hard and extra narrow beds, and plastic wood paneling.

"But all in all, these were the nicest and most comfortable accommodations that I'd ever seen in a camp, and I had lived in plenty," he said. "About the only creature comfort that wasn't properly taken care of was the dryness of the air."

One reporter who visited the so-called British Petroleum Hilton at Prudhoe Bay, with its swimming pool and imported fir and birch trees, called it the "Ultimate Alaskan House Trailer."

"The atmosphere of the place is a cross between college dormitory during semester break and starship bound on a generations-long colonization flight," he wrote. On one floor there were college kids working for the summer who were smoking dope and listening to Joni Mitchell on a stereo. On another floor, six veteran roughnecks dealt hands of poker with $700 in the pot.

Kate Wedemeyer, the recreation director at Dietrich Camp, said that camp was different from the others because it was in the mountains, and the scenery outside the pre-fab housing was spectacular.

"I think it really affects the morale," she said. "They say it's friendly, but I think that's partly because people are happy because of how pretty it is."

The ratio of men to women was about 10 to 1, which changed the social atmosphere from the routine of normal settlements where the percentages are not so skewed.

"For me I remember the first five weeks was really hard because people were making comments and checking me out to see where they stood and what they could get away with and if I had a boyfriend or anything like that, just generally bugging me. It really sort of got to me the first five weeks, but then they sort of found out what kind of person you are, and the word kind of spread," she said.

She said one of the worst things about life in the camps was that "people go through your life so fast up here." People often quit less than eight weeks after arriving, so much so that "two months is a fairly long time around here."

The biggest difference between the pipeline and previous construction jobs was the presence of women in the camps. Most of the time men and women had separate rooms, but on many occasions the men and women shared the same bathrooms and there were times when notices were posted in the camps saying "No one can be assured of receiving a room assignment with a person of the same sex."

Two workers at Old Man explained their views on the tensions created by the imbalance between the sexes in the isolated camps.

"Camp life seems to reduce some of the most independent rational adults back into high school, game-playing adolescents," said Judy, a rig oiler. "I find it's nearly impossible for me, too, to not play along with little ritual flirtations."

Bill, a crane operator, agreed. "I haven't acted or thought this way in ten or fifteen years, since I was a teenager. I thought that once you outgrew those patterns, that was it. But I'm right in there, competing with other guys to smile and talk with the girls. It just becomes important to be personally recognized, somehow singled out by someone, especially a lady, from all the other guys."

The women's part of this, Judy said, was the "huge mountain of attention looming in front of you every way you turn.

"Even the smallest compliments carry sexual overtones, maybe only because of the unequal number of men and women, but after a while, I just get so I want to go completely unnoticed as a female and be accepted simply as a worker," she said.

Making the time go faster and easier was a prime goal of all the official and unofficial activities in camps. "I have my cross-country skis up here and in my off hours I go cross-country skiing," said one woman. Another, Patty Smith, said she liked to spend time in her room, which was decorated with posters and pictures. She'd listen to tapes, watch TV, or have friends over.

Organized activities included pool and foosball tournaments, volleyball and softball games, and a variety of boat races on different creeks. In July 1974, for example, the "First Annual Dietrich Camp Float Race" flowed along a seven-mile course to the main Koyukuk River bridge.

All the entrants had to have boats that had rubber inner tubes in them. The "Vis-Queen Special," piloted by Todd Bennet and Bart Garber, won.

Because they were not able to participate in the 1976 Winter Olympics at Innsbruck, workers at Galbraith had their own Olympic competitions in the sports of Ping-Pong, foosball, shuffleboard, pool, darts, and arm wrestling, featuring competitors from Galbraith and Pump Station No. 4.

Gini Derrington, who worked in transportation/air operations, would light incense and lead a nightly yoga class, an activity that had just boomed into popularity in big cities and small towns across the country. Sitting cross-legged in her sweat suit, she explained that the class started with some friends getting together and it was an ideal way for them to unwind.

"You have a really lot of crazy activities during the day that sort of begin to get to you. The phones ringing. The people hollering. They get excited sometimes. So you come to yoga and you drag yourself in and you think, 'I don't want to do yoga.' But here I am because I committed myself and then we dance out. We just have a gas. It makes for a good evening."

Several of the camps had their own mimeographed newspapers: *The Atigun Times*, the *Old Man Kanuti Krier*, the *Coldfoot Chronicle*, *The Dietrich Digest*, *The Livengood Boondocker*, *Pump Station 9—On the Line*, and the *Toolik Tundraground*, formerly the *Toolik Tribune*. The Atigun paper called itself "America's Farthest North Daily Newspaper," while the Toolik sheet, which was farther north, said it was "Indisputably America's Farthest North Daily Newspaper."

Alyeska's own newspaper for the workers, *The Campfollower*, was heavy on pipeline poetry and feature stories about antics in the camps.

The camps had an average of five "morale circuits" on which residents could make phone calls. There were usually long lines of people waiting to use the phone, which made it hard for many people to keep in touch with their loved ones. "If you spent too long on the phone someone would start banging on the booth for you to hurry up," said welder Jack Yuers of California.

Especially for married men with children, the separation could be stressful. Tim Dove said he spent $100 a month to call his wife in Anchorage from Dietrich.

"Life isn't unbearable here," said Roy Deasis, a twenty-nine-year-old Native Alaskan working at Dietrich. "It's just that I'd rather be doing something else."

Henry Kissinger donned Arctic gear for his tour
of the pipeline in December 1975.

HENRY KISSINGER'S DREAM

The pipeline was like an eight-hundred-mile magnet for celebrities and politicians. Presidents, kings, senators, business executives, actors, writers, singers, and others included the pipeline in their travel plans.

Such was the popularity of a pipeline side trip that the decision by folk singer Pete Seeger *not* to visit the pipeline when he came north was deemed newsworthy by an Anchorage columnist.

Some of those who visited the pipeline included actor Jamie Farr, who played Corporal Klinger on the TV show *M*A*S*H*, singer John Denver, who deemed it a "far out" experience, comedian Arte Johnson, radio personality Arthur Godfrey, and singer Gladys Knight, minus the Pips.

Knight came to Valdez to make her screen debut in the first of two bad movies made about the pipeline, *Pipe Dreams*. In the movie, Gladys plays a wholesome Georgia teacher who heads north to Alaska not to make money, but to get her man back. He is played by her real husband Barry Hankerson, who had become a pipeline pilot. His dream is to retire in Nome, but Knight wants to get him back in Atlanta. After doing battle with the evil boss of vice in Valdez, Gladys emerges triumphant. She and her husband reconsummate their marriage with a romantic encounter inside a piece of pipe in the Valdez pipeyard.

Critic David Shear of the *Philadelphia Bulletin* said the director should "lose his license for making *Pipe Dreams*, one of the worst films ever made."

The other movie made about the pipeline was *Joyride*, another bomb, which wasn't even filmed in Alaska. In this movie, filmed in part in Roslyn, Washington, the town that would years later be the site of the TV show *Northern Exposure*, the search by young people for pipeline jobs ends in robbery, kidnapping, and a car chase. The main attraction of the movie was that it starred the offspring of four Hollywood legends—Desi Arnaz, Jr., Bobby Caradine, Melanie Griffith, and Anne Lockhart.

The best tongue-in-cheek line about the pipeline was spoken not by any of the actors in those movies, but by Secretary of State Henry Kissinger, who, after seeing the pipeline near Fairbanks, said: "When I was a little boy, it was my life's ambition to visit a real pipeline. Now I have realized my life's dream."

The stout Kissinger, who needed help from an airman in putting on his mukluks, added that he would "have more quotes for you when we get to the Great Wall."

Kissinger was in Alaska accompanying President Gerald Ford, who was on his way to China for a meeting with Mao.

About five thousand people from the Fairbanks area crammed into a hangar at Eielson Air Force Base to see President Ford arrive in Air Force One. It was ten degrees when the north end of the hangar opened and the gleaming blue-and-white jet pulled inside.

Ford's seventeen-minute speech drew boos at one point and cheers at another by different factions in the crowd. When he said that by any standard the benefits of the pipeline outweighed the disadvantages, the president was booed by a small group in the crowd who carried a banner that said "Please Save Us from the Pipeline $$."

Then he was cheered a few seconds later when he said the "pipeline has proven to be an outstanding example of how our ecology can be preserved while energy needs are being met."

Among the two hundred reporters in the presidential entourage was John Chancellor of NBC. "I'm trying to get to that god-damn pipeline," said Chancellor. "That's a good story."

While touring a section of the pipeline, the president talked to sideboom operator Roy "Hamburger" Vandiver. Ford, assuming Vandiver was from Texas, asked if he liked the Alaska weather better than the weather in Texas. Vandiver said he assumed the weather was better in Texas, but he did not mention that he was actually from Hornsby, Tennessee.

The idea that the pipeline was a "good story" also attracted, on another occasion, novelist Ernest K. Gann, author of *The High and the Mighty*.

He scouted the state with the idea of writing a script for a film about modern Alaska, but he figured it would be a tough challenge. He said that the industrial complex at Prudhoe Bay was "exciting, but as far as shooting pictures there, it might as well be Keokuk, Iowa.

"It's like someone says there's a wonderful story to World War II. It's a big story, fifty thousand stories. I don't know whether the story of modern Alaska is oil, or gold, or opportunism, or hookers, or what."

Sitting in a hotel coffee shop in Fairbanks, Gann said he was shocked to see Alaska so developed. Looking out the window of the Travelers Inn, he said, "We might as well be in Kansas City right now."

In addition to attracting artists and entertainers, the pipeline also proved to be a royal attraction. King Olav of Norway toured Prudhoe Bay as part of a trip to the United States marking the 150th anniversary of the first Norwegian immigration to the United States.

The wind chill factor was fifty below zero when he went inside to lunch at the Arco camp, taking tray in hand and going through the lunch line for a thick steak. Later he toured the area by bus, but the windows were frosted over, so the king dug out a credit card and used it to scrape the window.

Now that's something you won't see in Keokuk, Iowa.

Construction work continued in the deepest cold and darkest winter days with towers of lights to aid vision.

CAMP RULES

The "don'ts" of life in the camp, according to the official rules, covered just about every wrong imaginable.

Camp rules, some of them enforced and some of them not, were published by Alyeska for those living and working in the pipeline camps.

The list featured twenty-two camp no-nos:

- stealing
- unauthorized absence or lateness
- loafing on the job
- falsifying any reports or records
- making false injury claims
- leaving work without approval
- smoking in undesignated areas
- drinking alcohol
- possessing firearms
- abusing tools or equipment
- horseplay
- insubordinate conduct or refusing to follow foreman's orders
- gambling
- interfering with fellow employees
- refusing to accept work assignments
- fighting or violence
- dishonesty or fraud
- sleeping on the job
- failure to use safety equipment or wear prescribed Arctic clothing
- unauthorized use of transistor radios
- illegal use of drugs
- violating safety rules

There was also Rule No. 23, the violation of which, just like the violation of any of the other twenty-two, would leave employees "subject to disciplinary action including discharge."

Rule 23 prohibited "violation of any of the Camp Rules."

Under the camp rules, the music had to be turned down between 10:00 p.m. and 6:00 a.m., and "Violation of ordinary standards of cleanliness is prohibited," a rule that was open to debate regarding ordinary standards at construction camps.

The people in the camps were not allowed to sell anything, use the laundry after 11 p.m., have pets, be intoxicated, feed animals, or entertain unauthorized guests.

Sick workers had to report to the medical unit at 8:00 a.m., and fishing and hunting were not allowed along the pipeline route.

"The Contractor reserves the right from time to time to make such additional rules that may be necessary or required," the camp rules stated.

There were also general safety rules requiring hard hats in certain jobs, and one rule that advised, "You are cautioned not to run on this project, except in case of emergency."

There also was a rule against using compressed air to clean yourself or another employee. One other prohibition was that against throwing tools or materials down from high places.

Every worker received a copy of *Staying Alive in the Arctic*, which had tips on survival in the cold.

"Carry this booklet with you whenever you leave camp," it said. "Even if you know everything written herein, its pages can be valuable tinder for a fire."

"The pièce-de-résistance is at Prudhoe Bay: chocolate chip cookies as thick as crockery platters, as round as pancakes, and as bumpy as dozens of miniature cocoa boulders can make them."

— From a 1976 Associated Press report on food in the Trans-Alaska Pipeline camps.

ALL YOU CAN EAT

Like all other armies, the pipeline work force traveled on its stomach. And on the pipeline, the battle of the bulge was fought all over again. Meal time was the biggest break in the monotony of the day, and the lavish spreads in the dining halls became world famous.

Meals of steak, frog legs, Beef Wellington, lobster, and prime rib whetted tens of thousands of appetites and led to widely circulated news stories about extravagant menus that became part of the pipeline legend.

"The term 'lunch' had to be the greatest understatement of the day in a camp where even the noon meal is something akin to a Thanksgiving feast," Kenn Oberrecht wrote of a visit to Coldfoot in 1974, when the day's lunch special was king crab.

That same year another writer said the "food approaches fantasy, from omelets made to order any morning, through lobster, steak, and cake anytime, to flown-in fruit late at night."

In the first season of pipeline construction, a typical dinner started with the soup du jour, followed by New York steaks, Shrimp Louie, garlic bread, French bread, brussels sprouts with cheese sauce, fries, seven salads, peach pie, ice cream, and cookies.

Food was a morale booster, a direct way of making camp living more palatable. Since pipeline jobs for those on the union "A" lists, the workers with the first crack at jobs, were easy to come by, there were workers who quit jobs if they didn't like the quality of the steaks or the Roquefort dressing.

Because other elements of normal life were missing, the pipeline workers thought a lot about food. Within a few months, it wasn't unusual for camp workers to put on twenty pounds by overindulging at the chow line.

"People up here go to meals not only to be fed, but to feed their sense of pleasure," said Cheri McDevitt, a twenty-eight-year-old Californian working at Pump Station 5. "At least for most people, eating breakfast, lunch, and dinner gives them a feeling of subtle sensuality—just about the only sensual act they can perform in a place like this."

The boarding room ambiance of the dining halls never changed, but the menus did. The system of paying caterers for all their expenses plus a profit, the so-called "cost plus" approach, proved to be very expensive, so Alyeska switched to paying the five main caterers a flat fee per person. That led to moves to economize.

As one official put it, "Now that the operation is not cost-plus, we're all dealing with the world of reality: How much and what should a person be fed a day."

With the new limits, the extravagant touches were removed from the daily diet. And Alyeska cut back on what one official called the "out-of-control abuse of box lunches, extra beverages and donuts, which is now costing $8,000 per day."

There was no accountability on the use of box lunches, said the catering official, leading him to ask, "Who else in Fairbanks are we feeding?"

People were walking off with cases of soda along the pipeline, adding $1,000 a day to the cost of the project. Extra donuts going out the door were costing $400 a day. Some people complained when Alyeska imposed a two-sodas-per-person limit for lunch and clamped down on take-out donuts.

Even with the leaner budget for catering, the all-you-can-eat daily pipeline diet for many ran from 4,000 to 8,000 calories and there was steak once a week and better food than most workers had at home. The average worker was downing a total of nearly six pounds per day of meat, fruit, milk, and vegetables.

"Everybody's overweight—so they're doing all right," said Chandalar cook Wooster Campbell, a retired Air Force cook who had also worked in the Captain Cook Hotel in Anchorage.

A typical camp breakfast may have featured Canadian bacon, beef patties, broiled bacon, and eggs, soup, hot and cold cereals, juice, fruit, French toast, pancakes, baked cakes, and pastries. In one camp the French toast was known as shingles, pancakes were called Frisbees, and hash browns were called sawdust.

When the cutbacks came, in the words of one worker, the fabled meals were replaced by endless lines "ending at overdone roast beef or dried-up calves' liver."

From then on, the opinions about pipeline chow fell into one of two opposing camps. On one side were those who were thrilled that they didn't have to pay for their meals and recognized that they didn't eat like this in the real world. On the other side were those who said the food was terrible. It was a tradition for people in the military and construction camps to complain about food, and in this respect many pipeliners were traditionalists.

"There's nothing to really look forward to except mail and food," one worker complained. "You get so you look forward to the meals, you think about them. So when they screw around with your food, they're really screwing around with your mind."

Another worker complained about cold biscuits, brownies that were like "chocolate-flavored tar," pastries that were "hard enough to drive nails," and lukewarm dinners.

A culinary worker who thought the food was fine said the protesters "remind me of a bunch of seagulls who eat, sleep, and squawk. They never had it so good in their lives, making a fabulous salary, free room, three meals plus a night-time snack."

He said one way to solve the complaints would be to have Alyeska hire Vietnamese refugees. "Grow up and act like men and thank your gods that you're living in a country that gives you this opportunity to work or be on welfare if you so desire," he said.

Chaplain Ray Dexter, reflecting on the opposing opinions, said he often stopped to wonder that such divergent views could come from people eating the same food.

"The old-timers who wistfully remember the halcyon days of last year with steak and lobster and crab salad, etc., etc., have some reason to grumble," he said in 1975. "Others who come up after several seasons of cooking for themselves out of cans and TV dinners are enthusiastic. We all eat the same food, but our reaction to it is colored by things other than the quality of the food itself."

Dr. Phil Nice helped establish the medical clinics in the camps, and he said the eating habits of workers gave him cause for concern. In at least one camp dining hall, there was a warning over the serving line about the dangers of overeating and heart disease.

"There was too much good food, so almost everybody gained weight, and obesity was a problem," Nice said. "The hard work was done by machines. Eating became a big thing on the line, and there was food available between meals. The meals were better meals than most of us can afford to buy. That, we felt, was a problem, but it was difficult to do anything about it."

David McCracken, a mechanic from Montana, told of following an operator in the food line who said to the cook that he wanted his steak "Rare, times two."

The operator returned to his table with two pounds of rare steak on his plate. The worker cleaned his plate, but he was an exception, McCracken said.

With two main meat dishes offered most of the time, there was "inexcusable waste," McCracken said.

There was fresh milk and more varied offerings than those of many restaurants, but still some people complained.

"It was gratifying to hear the food praised on occasion," he said. "A little exchange in the chow line made me smile. Somebody had sounded off about the 'slop in this camp,' whereupon the guy in line in front of him turned around and said, 'You know, if you ate this well at home, you'd have to earn this well to pay for it, and if you earned this well at home, whatcha here for?'"

McCracken believed that some of the food complaints came from the worker about ready to quit or "drag up," who worked himself up into an irrational frenzy for a reason.

"Psychologically, it seemed he had to justify the foolhardiness of whiffing such a well-paying job," McCracken said.

Complaints about the food often had more to do with what was on the minds of the workers as opposed to what was

Marcus Halevi

In the beginning, the pipeline boasted rich food and extravagant meals. With budget tightening, kitchen crews had to find more economic ways to feed the troops, angering some workers. Mealtimes became excessively important to people who had few other highlights to look forward to each day.

on their plates. Said one catering official: "They can't fight with their wives because they're not there, and they can't fight with the foreman, or they might lose their jobs, so they fight with the cook. If the camp runs out of, say, mustard, you'd better fly it from Seattle or Chicago or there could be a riot."

Sixty-nine members of the pipeyard crew in Fairbanks wrote a letter of protest to the pipeline newspaper about the sack lunches, piling on the adjectives like layers on a hero sandwich: "We cannot tolerate mung-meat sandwiches, rancid tomato slices, tooth-breaker cookies, fried dried pork chops, watery fried chicken, and generally sorry foods packed the night before and left to ripen in a warm room before we get them the next day."

Twenty-one dissatisfied workers, who signed themselves as "The Hungry Horde" of Galbraith, said their food was the worst north of the Yukon River and what was tossed in the garbage made for the skinniest ravens on the North Slope.

"Not only is the food of extremely poor quality, it is not served hot," the Horde said. "The cooks refuse to put water in the steam table, and laugh off the suggestion of having a heat lamp installed over the dinner roll tray."

Gene Lyda of Alyeska said there was some truth in the letters from the pipeyard crew and the Horde, but "in some cases they obviously reflect a known group of employees led by agitators."

He said that the Alyeska public relations arm should counteract the work of the agitators by putting out positive stories about catering in the camps. Trying to satisfy a group of people whose tastes ran from grits to granola, and burgers to black-eyed peas, was a difficult job.

One attempt to improve the atmosphere was the special theme meals offered by some of the camps, like "Dinner at Diamond Head," an evening when the workers in Carhartts and T-shirts took a taste trip to Hawaii, with dishes ranging from macadamia nuts to Curried Chicken Polynesian. The Hawaiian dinners brought thoughts of sun and sand to workers in winter.

An immense logistical effort was required to supply the camps. Greyhound Support Services, one of the contractors, said at a typical three-hundred-man camp, the weekly requirements were as follows: 480 gallons of milk, 800 pounds of steak, 300 pounds of lobster tail, 60 gallons of ice cream, 210 pies, 910 dozen cookies, 458 dozen eggs, and 1,000 pounds of hamburger.

On some occasions, complaints about food went far beyond mere letters of complaint or idle talk about the cook's shortcomings. Workers dumped loaded food trays on the floor, spilled the silverware containers, or tried to organize food boycotts, but hunger pangs usually brought them back to the table without missing too many meals.

The boycotters griped about stale sandwiches, getting steak only once a week, getting prime rib only every other Sunday, the lack of supplies at the salad table, the lack of condiments, and food that didn't look appetizing.

Mike McDermid, camp manager at Dietrich, said he had seen everything from a "toothpick rumble to a buttermilk caper" arising from the mess hall.

"If there aren't any toothpicks, we have a toothpick rumble. The next day there may be toothpicks, but we'll be out of buttermilk, so they have a buttermilk caper," he said. "And there had better always be clean towels and steak sauce. Most of these pipeliners have one standard reaction when something goes wrong—'I'm gonna tear this place apart.'

"It is a constant threat simply because they do not know any other way to vent their frustrations," said McDermid.

One of the most serious disputes took place at Tonsina Camp near the southern end of the pipeline when twenty to thirty members of the Local 798 pipeline welders union rioted at 6:00 a.m. one June day, injuring two food service workers and causing $3,000 damage to kitchen facilities. They began overturning tables and damaging the serving line equipment after being denied raw steaks to take with them to the field and barbecue for lunch. The welders had made it a habit to raid the kitchen and the freezer before going to work, taking pies, other food, and steaks that they broiled on makeshift barbecues.

That day they had been asked to take sack lunches of sandwiches, fruit, and dessert, but such was their attitude that two Alaska State Troopers had to be sent to the camp to maintain order. Several kitchen workers were injured in the melee, including Willie "Babe" Newton of Anchorage, who suffered a painful leg bruise from a flying table.

"Enough food was going out of that camp to feed three hundred people at lunch," said Tom Adams, a spokesman for the Culinary Workers Union. "I'm surprised the environmental people hadn't complained about the fires out on the line. It probably started small and then grew."

The spark that touched off the confrontation was that culinary workers had complained that if steaks were getting cooked at lunch, Culinary Union workers should have been doing the cooking, not welders.

The jurisdictional beef was settled when the contractor announced that workers would be allowed to take the precooked "meat of the day" with them to the job site.

The Tonsina food fight struck those in Alaska who were not working on the pipeline as a prime example of the excesses brought by the era of easy money.

Anchorage resident Jean von Dohrmann said the men were acting like "spoiled brats."

"If they were mine they'd get the seat of their pants warmed, and if they were too big for that, then the cost of repairs would come out of their allowance," she said. "The whole episode is disgusting—especially to those of us who have seen what a forest fire can do to Alaska and have had to scrimp for two weeks to manage a steak dinner and who watch the pipeliners with their ninety-nine tax exemptions rip off Alaska.

"Here's hoping Alyeska has a big enough stick to make the culprits sit in a corner for their tantrums," she said.

A Memorable Menu

Thanksgiving Dinner

Franklin Bluffs, November 1975

Soup
French Onion with Croutons

Appetizer
Seafood Salad in Remoulade

Entree
Roast Turkey with Souffle Dressing
Old-Fashioned Giblet Gravy
Cranberry Sauce
Glazed Virginia Baked Ham with Romaine Sauce

Vegetables
Mashed Potatoes
Candied Potatoes with Marshmallow Topping
Brussel Sprouts with Cheese Sauce
Creamed Corn with Butter Croutons

Salads
Cranberry Mold
Tossed Green Salad
Coleslaw
Asparagus Vinaigrette
Fruit-Fluff
Pickled Beets
Cottage Cheese
Assorted Relishes

Dessert
Thanksgiving Cake
Pumpkin Pie
Mincemeat Pie
Banana Bread

Breads
Assorted Dinner Rolls
Cornbread

Beverages
Milk, Tea, Coffee, Hot Chocolate, Assorted Juices

THE NAKED TRUTH

Along with mood rings, pet rocks, and other fads of the mid-1970s, streakers blazed their way through the pipeline camps. The practice of running naked through public places began on college campuses and had its own anthem, *The Streak*, by Ray Stevens.

On the pipeline, the first streakers to be photographed were "The Magnificent Seven," a group of men who ran through Atigun Camp, the "Shangri-La of the Trans-Alaska Pipeline," on December 18, 1974.

Shouting "Merry Christmas!" and "Ho-Ho-Ho!" the men followed a route that took them through the warehouse, light truck park, Arctic walkways, and the theater. All this on a thirty-two-below-zero night.

At Galbraith Lake, streakers were required to follow pipeline regulations. There was a streaking roster for the different trades, and prizes for "Streaker of the Month," such as a one-week vacation at Prudhoe Bay.

"In a nod to project safety requirements, all streakers are being required to wear bunny boots and hard hats," the *Campfollower* newspaper reported.

NEW BEGINNINGS IN THE PIPELINE PARISH

It was a wedding on the pipeline—literally. The bride, Katherine Cline, had purchased a bridal gown and high heels, but thought it best not to wear them while standing atop the forty-eight-inch pipe about eight-four miles north of Valdez along the Richardson Highway.

She was dressed in a green pants suit and a "Prudhoe Bay to Valdez" jacket. The three hundred people who attended the ceremony wore hard hats.

The Rev. Roger Weaver of Glennallen stood precariously atop a vertical support member, one of the pilings that supported the elevated pipe, while he read from the Bible. He addressed the five-member wedding party that stood atop the pipe.

Next to Cline was the groom, Jack Fike, along with best man David Veal, matron of honor Mrs. Tompie Newberry, and her husband. Cline, a widow and a grandmother, was living in Valdez, while Fike was a blasting foreman for MK-Rivers who had lived in Alaska for twenty-six years.

Later that year, the first wedding in a pipeline camp of two project employees took place at Coldfoot, where Caterpillar mechanic Tom Ramero and culinary worker Gail Delaney exchanged vows in Warehouse B. Their three-layer wedding cake featured a sideboom and a bulldozer.

At another camp wedding, this one at Tonsina, Frank Catone and Gayle Woodrome walked down the aisle while a phonograph played "We've Only Just Begun." They considered Tonsina their first home, not their second. "It's a small, self-contained city. You're going to have problems here like you would in any settlement . . . that's just the way it is," Catone said.

Chaplain Ray Dexter said that many of the thirty-two volunteer chaplains who served on the pipeline performed weddings. He doesn't know how many of the marriages lasted.

"I would say they probably had as good a chance as any. At least they had shared a common experience," he said.

Dexter, who had served for twenty years with the Salvation Army, coordinated and recruited the volunteer pipeline ministers, working out of an office that was upstairs from Tommy's Elbow Room, a popular bar on Second Avenue in Fairbanks.

Dexter said that work on the pipeline ended some marriages and strengthened others. The strains of separation led to numerous divorces, but many marriages were helped because pipeline wages brought a measure of financial security to many who had never had it before.

The chaplains' program was funded by Alyeska upon a suggestion from the Fairbanks Social Concerns Task Force of the Alaska Christian Conference. The biggest problem for the chaplains was that many of the workers were lonely, even though they were in close quarters with hundreds of other people.

"Our chapel program is the one place where Teamsters, welders, and Alyeska people can get together on a common plane," said Dexter. "It's the one place where they can sit around and relate to each other as people."

The chaplains were like circuit riders, traveling from camp to camp. Because the work continued seven days a week, "Sunday" arrived whenever the preacher did. Father Jack Gurr, a Jesuit priest, visited camps north of the Yukon, at services that "God, weather, and Wien made possible." The latter was a reference to the Wien Air Alaska F-27s that Father Gurr traveled in to the camps. He enjoyed the people he met in the "Pipeline Parish."

On one occasion, the priest was given a room in a camp that was adorned with nude pictures of women. "I didn't want to destroy someone else's property, so I just tacked up some pages from the *News-Miner* over them while I was there," Father Gurr said.

"I think a lot of people get some weird ideas about the type of people working on the pipeline, because most of the time it's the troublemakers that you read about in the headlines," he said.

Father Gurr was in one camp when several workers who didn't like the food that day filled their trays and dumped them on the floor. In his sermon, he said there were other ways to make their grievances be heard.

Among the other ministers were Sister Ellie Brown and Sister Alice Legault, who found that a big part of their work was to listen. "It would blow people's minds when they met us and discovered we were not ordained ministers, but women and Sisters, and that we were at the camp for something else besides money," said Sister Alice. "These people needed to talk and not just about God, but about their families, and why they were there. Some were experiencing real conflicts."

All major Christian denominations were represented by the chaplains, who were provided room and board, and transportation to the camps, but no salary.

The chaplains who traveled up and down the pipeline also were called upon to perform memorial services and baptisms. The first baptism by submersion took place in the Happy Valley gym, where shop foreman Otis George rigged up a six-foot-long and three-foot-deep tank for the baptism of lowboy driver Tom Tessier.

"It was overwhelming emotionally," Tessier said after being dipped by the Reverend Lindsay Williams, a Southern Baptist. "When I got out of the tank, I cried. That was all I could do."

TUNE IN NEXT TIME

In these days when scarcely a place on the planet is beyond the reach of live satellite broadcasts of CNN newscasts or Chicago Cubs games, it is worth noting the arrangements required to supply TV and movies for the pipeline camps.

There was no channel surfing. Tapes of television shows and reels of film had to be shipped by air for delivery to camps up and down the line.

The films, a different one each day, were provided on a regular camp-to-camp circuit and shown once in the evening for the day shift, and again in the morning for the night shift. Workers who helped out by running the projector in one camp received cards identifying the holder as a "certified exhibitionist."

At Old Man Camp, there was a theater with seating for 180 and quadraphonic sound. Movies at the camps were, in many cases, more recent than those shown in Anchorage and Fairbanks. Action films drew the biggest crowds.

Jaws, the oceanic epic that made Americans afraid of going into the water at the beach, appeared in May 1975 on the pipeline, a full year before the shark saga appeared on the silver screen in Fairbanks.

On television, when Walter Cronkite ended his CBS news broadcasts with the catch phrase "That's the way it is," it should have been amended to "That's the way it was two days ago."

TV broadcasts in the pipeline camps started in 1974 with taped syndicated shows and two-day-old national news flown in from Fairbanks and Anchorage, where it was broadcast as day-old news. In those days, the comedies and dramatic shows carried by the networks were shown three weeks late in Fairbanks, after first airing in Hawaii and Anchorage.

Taped shows in the pipeline camps included *All in the Family, M*A*S*H, 60 Minutes, Cannon,* and *The Six Million Dollar Man.*

Timely broadcasting took a step forward when part of the 1974 World Series between the Los Angeles Dodgers and the Oakland A's was seen in the pipeline camps on a one-day delay. An Anchorage ad agency, seven advertisers, RCA Alaska Communications, and NBC had arranged that year for the first live satellite telecasts in Anchorage and Fairbanks of the fall classic. It cost the sponsors $5,000 per game.

The deal came together only twenty-four hours before Game 1, but the TV stations in Alaska still didn't have permission from Baseball Commissioner Bowie Kuhn to tape the games for viewing in the pipeline camps.

Finally, five hours before Game 3 of the series, the Alaska promoters received the official okay from the commissioner's office. The balance of the series, which the A's won for the third year in a row, was taped at KFAR in Fairbanks. The tapes of the games were flown to eleven camps within twenty-four hours of the last out each day.

Alyeska had installed nineteen-inch TV sets in the recreation halls of the camps in September 1974, showing tapes of such programs as *Voyage to the Bottom of the Sea* and *Daniel Boone*.

By the end of that year, workers could watch TV in their own rooms if they brought a TV with them to camp. With approval by the Federal Communications Commission, tiny ten-watt camp transmitters were set up in each camp. The stations were on the air four hours at night and four hours in the morning, at the end of the two daily shifts.

Ads promoting pipeline safety were inserted into the broadcasts, including one by Medical Director Dr. David Livermore, in which he pointed out the dangers of sneaking a bottle of liquor and hiding it outdoors: "When the weather gets very, very cold, below zero, if you drink that alcohol straight out of the bottle, it can be like drinking lye, and you don't want to spend months having a transplanted esophagus made out of your own intestine, would you?"

Jim McMillian, a camp official at Prospect, said television quickly became very important for workers.

"After a long day of working outside, a lot of people don't like to be crowded into the recreation hall to watch a movie.

They much prefer to be alone. To these folks, being able to watch TV makes even a camp dormitory room a little spot of home," he said.

TV Guide published an article on the TV arrangements by Alyeska, and quoted radio operator Cher Lichner as saying she had seen so many TV detective shows on the pipeline, she was becoming an expert on the genre.

"Her first name brought the attractive young lady in for no small amount of kidding after the 'Cher' special this past winter," *TV Guide* said.

"The guys all said that all the other Cher and I had in common was the name," Lichner said.

Also broadcast in the camps were televised courses in log-cabin building and how to save on income taxes.

"Retiring to a cabin in the woods is still a lively dream up here. As for income taxes, the men just want to know how much money they get to keep," said Bill White, an employee of the recreation office.

LETTERS HOME

The Livengood Boondocker, which billed itself as the bi-weekly answer to the *Wall Street Journal*, had a remedy for those who found themselves too tired to write home after a twelve-hour day.

A semi-form letter was appropriate for all occasions to inform friends and family about life on the pipeline. All the writer had to do was to take the following semi-form letter, circle the appropriate choices, and mail it home.

Dear Wife, Parents, Children, Friend, Parole Officer, *(circle one)*

I am now working at *(circle one)* Tonsina, Chandalar, Coldfoot, I don't know, and am having a *(circle one)* great, fair, average, lousy time.

My roommate is a *(circle one)* bear, Teamster, caribou and he *(circle one)* drinks, swears, all of the above.

It is a rough life up here but don't worry about me. I haven't been in a fight *(circle one)* yet, since last month, this week, since breakfast.

We are all working hard and we will get this pipeline built *(circle one)* soon, next year, the year after, God knows when.

It will be winter soon and I will be *(circle one)* cold, frozen, in Hawaii.

Am looking forward to hearing from you.

See you
(circle one)
Soon, Someday, Maybe.

FARTHEST NORTH GATOR

While there were plenty of bears on the Trans-Alaska Pipeline, there was only one alligator. He was in the possession of Marty McCasland, a welders' helper at Galbraith Lake.

McCasland, who played the harmonica and was the official pipe gang bugler, was known for being able to swallow an apple or a banana with one bite, which led to discussion about whether he could swallow a compressed sixty-pound bale of hay.

A 798 welder who came back from Louisiana on R&R had returned with the eight-inch animal in a Crest toothpaste box, telling McCasland, "By God, I brought you an alligator."

On the flight to Alaska, the welder occasionally had visited one of the plane's washrooms to dip the alligator in the sink.

"Well, I was doing some partying, and everybody wanted to see this alligator," McCasland said. "Half of them didn't believe it. We went down to the room and this one girl said, 'Oh man, does that bite?'

"I said, 'Heck, no, it won't bite.'" He grabbed it and kissed it. The alligator proved him a liar by biting McCasland on the lip.

"I said, 'I weigh 220 pounds and if you little eight-inch devil gonna bite me, I'll keep you,'" McCasland said. "One of the guys who used to be an old gator hunter called him Albert. That was his name: Albert Gator."

McCasland, who came to Alaska in 1967, said that at first all he heard was how he wouldn't be able to handle the 798ers.

"The more I heard about 798ers, I thought to myself, 'Hell, I'm going to like that group.'" He joined the union and became a welders' helper.

"I've lived happily ever after," he said. But Albert Gator didn't do as well.

"He ended up dying in Galbraith, and we had a full-fledged funeral for him," McCasland said. "Dixie flag, the whole pipe gang. There was about 152 people came to the funeral."

McCasland had three boxes of Cuban cigars, which were

distributed among the mourners by women who worked as welders' helpers. Four welders served as pallbearers.

"We took him out on the tundra," McCasland said. "One of the welders had welded a cross, put on it 'Albert Gator, 1976,' and just outside of Galbraith there we put him in the ground.

"Without a doubt it was one of the highlights, party-wise, of the pipeline," said McCasland, who spent $500 on his alligator's funeral. "It doesn't take much to amuse you up here."

Marc Olson/Fairbanks Daily News-Miner

With skill, pipeliners could make a few extra dollars on the side by crafting pipeline art from scrap pipe.

A PIECE OF THE PIPELINE

In a state where moose droppings are turned into objets d'art, it was only natural that the Trans-Alaska Pipeline would provide the raw material for souvenirs.

We're not talking here about the stray wrench in the bottom of the Alyeska parka, or sheets of blue foam insulation in the back of the pickup. We're talking pipeline maps, ashtrays, lighters, dishes, hats, games, beer mugs, T-shirts, pipeline socks, belt buckles with gold medallions, tie tacks, and necklaces.

The creation of Alaska maps, cut with torches from scrap pieces of the forty-eight-inch pipe, were as common on some job sites as pencil sketches in art class. The maps had the pipeline inlaid in brass from Prudhoe Bay to Valdez. A welder with exceptional expertise said one day his foreman shut down the job to allow the men to make maps.

Pipefitter Potter Wickware said it took skill to get Southeast Alaska and the islands of the Aleutian Chain just right. "Then you polish the surface of it with fine-grit sandpaper until it shines. Then you blue it with the torch, bringing out the various subtle shades that are inherent, but normally invisible, in the grain structure of the iron: violets, reds, blues, yellows. Some of the expertly made maps are really beautiful, and would stand in the same class as bibelots from the Fabergé workshops.

"If you have a large, finely made map, polished until it shines, tinted and inlaid, you can take it into Valdez any day of the week and get five hundred dollars for it," Wickware said.

At another camp a worker expected to clear $7,000 from his half of a two-man map-making team. The enterprise was threatened, however, after a controversy arose over a worker who had cut a map from a piece of pipe that was supposed to be shipped to the Lower 48 for tests. He lost his job.

Pipeline belt buckles were a big item, too, but like the maps, great care had to be exercised in their creation, because if you got caught creating one on the job, you'd get fired.

As far as store-bought mementos, workers on R&R were liable to buy almost any souvenir with a pipeline motif. Newland's Sourdough Gift Shop on Second Avenue in Fairbanks sold $50,000 worth of souvenirs in a two-month period in 1975.

"We're making a heck of a lot of money off it," said Manager Marvin Newland. "It's too bad they don't come out with some kind of pipeline shoe."

There were sweatshirts that featured Uncle Sam and an Arab character with the words "Who Needs You" directed at the latter. There were T-shirts for children with the words "Junior Alaskan Pipeline Worker."

The ultimate pipeline gift, sold by an outfit in Fort Worth, Texas, was the pipeline ring, which, in the words of the ads, was "Dedicated to the Chosen Few Who are Engaged in the Construction of our Nation's Largest Pipeline Project."

"First, to compliment the project itself, the ring must be massive," the ad said. "It was patterned after the style and size of the famous Super Bowl rings, and will create a sensation wherever it is seen."

The massive rings were personalized with the name of the worker and featured a map of Alaska showing the pipeline route. "This ring will last a lifetime and will serve as visual proof that you participated in the construction of the great pipeline. It is being offered only to those who are engaged in the work of the project. To sum it up, we know that you will wear it with pride. You will stand above the rest."

And like the pipeline, the pipeline ring was not cheap. The basic 10K model with no stone sold for $335, while the deluxe was $795 for a 10K ring with a half-carat diamond.

PATRIOTIC PIPELINERS

To celebrate the Bicentennial in 1976, 112 workers from eight camps chartered a jet and flew to Nome for three hours of fun in the Midnight Sun.

It was a pipeliner's Bicentennial. Wrap up a twelve-hour workday, hop aboard a jet for a one-hour flight to the historic gold town of Nome on the Bering Sea coast, and party for a few hours before riding the plane back to Prudhoe in time for work the next day. It cost them $160 each for the air fare.

"They said they had so much fun, they wanted to get another charter in six or eight weeks," one Nome resident said.

Ironworker Len Millard of International Falls, Minn., and his friend Ron Price organized the outing, made more difficult by the skittish airlines. One airline agreed to handle the charter, then backed out. A deal with a second airline was scrubbed on the night before the flight.

Millard made frantic calls to the first airline and finally got a plane, which lifted off about an hour and half later than originally scheduled at about 9:30 p.m. on the night before the Fourth of July Bicentennial.

Because Nome was in a different time zone, the workers arrived in Nome at about the same time they had left Deadhorse. Millard said he thought up the trip because he figured, "it would be nice to get out of here for awhile." He tested the concept a couple of weeks earlier by getting thirty people to go in on a one-night charter to Barrow to attend an Eskimo whaling festival for a few hours.

While the rest of the nation was celebrating with tall ships and fireworks, the pipeliners were standing in the airplane and enjoying the national anthem while crossing the Arctic Circle at twenty thousand feet.

"It was touching, very moving," Millard said. "We all stood up and played *The Star Spangled Banner* on our kazoos. It brought tears to my eyes."

A friend of Millard's from International Falls, Tom Reardon

was then in Nome operating the Breakers Bar. Reardon had buses at the airport to escort the pipeline delegation to downtown Nome, where several businesses stayed open all night to accommodate the group.

He also gave them the run of his bar. His only worry was that not everyone could fit in the bar at once.

After honoring the nation's birthday, the workers headed back to the airport, but the 3:30 a.m. departure was delayed by poor weather.

A break in the clouds finally allowed the plane to get airborne, so the workers made it back to Deadhorse, with enough time to catch buses for work—and maybe a little shut-eye.

MULESKINNER ON THE HAUL ROAD

Her legal name was Margaret Merriman, but to hundreds of truckers on the North Slope haul road, she was simply the "Muleskinner."

Her place wasn't a commercial truck stop, but a homesteader's cabin on a straight section of the Elliott Highway where truckers stopped at all hours to enjoy free homemade pie, coffee, and conversation.

Family pictures of the regular visitors adorned the cabin walls. Such was the traffic in and out of her front door that she made gallons of coffee with fresh spring water hauled from near the cabin.

The Muleskinner's homestead was always open, she said, because it was a practical way of putting her Christian beliefs into action. Call it a pie ministry for the pipeline's road warriors.

"A lot of them didn't understand it, but that was our work in the Lord," she said.

The Muleskinner and her husband, Howard, whose CB handle was "Thunder," had homesteaded in 1964 at Tolovana, about seventy miles north of Fairbanks, after migrating north from Florida.

While others built the pipeline, the Muleskinner built a formidable reputation as the unofficial hostess of Livengood. She had nine children and was able to keep them in line, thus the name Muleskinner.

Many people first heard of her after an encounter with a driver who had stopped on the road outside her house, she said. It seems the driver was heeding the call of nature in sight of her kids without bothering to go into the campground on the other side of the road. The Muleskinner went out and fired a warning shot in the air with her pistol.

"I said I'd amputate him if he didn't get his britches on," she said.

He did, so she didn't.

They later became friends, she said.

"I felt like the whole time up there I was either in an old novel or a comic strip," she said of her years in Alaska. In the late 1980s, several years after Howard's death, she returned to Tennessee.

The Muleskinner's unusual brand of pipeline mission work was financed by money that Howard made working as a powerhouse operator at the Livengood pipeline camp. The two of them gave away thousands of dollars worth of pie and coffee.

"They're very outstanding people in my eyes," said trucker Phil Menges in 1976, when he was better known as "Pappy Yocum."

"My own personal feeling is that well over half of the truckers up and down the road feel that they are a great asset to truckers just by being there. There's always a cup of coffee. It really shortens that long distance between places."

The Muleskinner's was the first stop for many truckers after they got out of the Fairbanks area and headed north.

Most of the drivers preferred apple, cherry, lemon meringue, or peach pie, but there were two drivers, known to her as "Black Beauty" and "Thunderbug," whose tastes ran to rhubarb pie.

Next to the Yukon Stove in the Merriman cabin, supplied with electricity through a portable generator, was their CB radio. Tuned to Channel 19 round-the-clock, it gave her access to the inner sanctum of trucking that passed by the front door.

She became as much a part of the haul road picture as "Sneaky Snake," the "Bootlegger," "Star Shine," and the seven hundred other truckers she knew by the CB handles she recorded in her notebooks.

She had started to learn about the CB when she heard one of the truckers say, "Breaker one-nine, here's the old Hoot Owl coming through. Does anybody have a copy on the Hoot Owl?"

Her husband had to explain that the Hoot Owl was trying to start a conversation with anyone who, in CB lingo, "had their ears on."

Muleskinner, then about fifty years old, started using the radio to relay emergency messages about accidents to Fairbanks, and she talked to the drivers as they passed by.

Whenever there was trouble on the road, the Muleskinner would help get word back to Fairbanks, relaying messages to "Wild Woman" on Tatalina Hill who in turn would call "Ridgerunner" near the Hilltop Cafe.

In other parts of the country and across Alaska, the CB radio's popularity was a fad that grew out of frustration with long lines at the gas pumps and the national 55 mph speed limit.

Millions of Americans had started to say things like "10-4," "put the hammer down," and "pedal to the metal," as they joined the CB fad. They took to the airwaves to talk to their "good buddies" with the same fervor displayed in the 1970s for the Pet Rock and Disco Fever.

Popularized by the hit song "Convoy," the CB boom helped create an image of the truck driver as a reincarnation of the American cowboy.

Keeping the cowboys' broadcast language in line on the haul road was one of the chores taken upon herself by the unofficial hostess of Livengood. If they were going to swear, they should be doing it without keying the mike, she said.

There were times when a truck driver would be out of line on the radio and the Muleskinner would "come in there and straighten that situation out," Menges said.

"She really does it beautifully," he said. "I've heard this several times myself. If there's any resistance at all, usually a trucker will come in and back her up. Usually it's not necessary."

They didn't mess with the Muleskinner.

SECTION III

SUDDEN AND UNEXPECTED WEALTH

A happy William Roderick drew a paycheck that included straight pay, time and a half, plus ten and one-half hours of double-time at $22.92 an hour.

"I want to save some money and let my son go to law school. It's not easy. I miss lots of things back home in Alabama, even things as little as good old Kentucky Fried Chicken. But I'll stick it out and smile good when my boy gets his degree."

— **Virginia Jackson, Isabel Pass camp cook, summer 1975**

SUDDEN AND UNEXPECTED WEALTH

At twenty-four, Teamster Greig Craft was making more money than he had ever thought possible—$1,100 a week for working "seven tens," seven days a week, ten hours a day on the Trans-Alaska Pipeline.

If paychecks piling up at an annual rate of $57,000 a year seem respectable today, they seemed enormous in 1975. Because of inflation, a worker would need to be earning more than triple that amount in the late 1990s to equal the buying power of $57,000 in the mid-1970s.

A lot of pipeline workers had tax returns that put them right up there with members of Congress, who earned $42,500 in 1975, or professional football players, who earned an average of $40,000. Many truck drivers surpassed the $50,000 salary of Alaska Governor Jay Hammond, the $60,000 salary of Secretary of State Henry Kissinger, and the $62,500 salary of Vice President Nelson Rockefeller.

This, at a time when gas was about 75 cents a gallon in Fairbanks, coffee was 25 cents a cup, and a new Chevy Nova could be had for about $3,500.

It wasn't the basic hourly wage rates that made for huge earnings on the pipeline, though those were mighty alluring to American workers beset by the worst economic downturn in a generation. The real bonanza was overtime, because there was time-and-a-half for every hour in the day after eight and every hour in a week after forty. Double time on holidays.

The most incredible series of paychecks went to Jerry Thornhill, a job steward for the welders union. From October 22, 1976, to November 17, 1976, he was paid for twenty-four hours a day, seven days a week. Then he worked three days at sixteen hours a day, before going back to twenty-four hours a day until December 5. In nine weeks, he made $35,344.01.

On that same crew, which was trying to complete work in Thompson Pass near Valdez, there were eighteen welders and welders' helpers who were paid for twenty to twenty-four hours a day for one week. Alyeska President Ed Patton defended the massive paychecks by saying the men were on the workpad all that time during pressure tests of the pipeline to make sure it didn't leak. Although they may have slept in the buses, the contract required that they be paid.

"Hell, everybody knows you can't work twenty-four hours a day," Thornhill later told reporter Bruce Morton of the CBS News broadcast *60 Minutes*, when his round-the-clock work habits were questioned. "But we were on call twenty-four hours a day. They were just giving money away."

In 1976, hourly pipeline union wages ranged from about $11 an hour to more than $18 an hour, depending on the job. With normal work weeks of seventy to eighty-four hours, people were making two or three times more than they had ever earned before because of time-and-a-half. Craft, like many other Teamsters, planned to take off twelve weeks a year, which would hold down his income to $44,000.

To help manage his extraordinary income, Craft wrote to *Money* magazine for advice, commenting that "sudden and unexpected wealth has left me somewhat dazed."

His uncle, who owned a trucking business in Alaska, told him about money to be made in Alaska. "When I heard the figure $6 an hour while I was making $2.25 an hour, it sounded like the end of the rainbow," said Craft, the son of an Army colonel, who had worked construction during summers while attending the University of Arizona.

Craft, described as having "an engaging smile that glints through a luxuriant brown beard," joined the rush to Alaska, signed up with the Teamsters Union, and went to work in warehouses in Fairbanks and points north.

He bought a new Toyota, a used mobile home, and two and one-half acres of land in Fairbanks, using money from his first year on the pipeline to get established in Alaska. Like many pipeline workers, Craft had more ideas than he had dollars in his pocket. He also knew that the big money would only last for a few years.

At first, Craft hoped to build a villa in Mexico. Then he thought of opening his own storage company, a truck-leasing business, or a hair-styling salon with his girlfriend. He also considered building an A-frame house on Chena Ridge, just west of Fairbanks, and constructing a vacation home in the Southwest. The pipeline had opened up a new world of possibilities, and he asked for the magazine's help in sorting through the options.

Money's editors agreed, but first they had to find financial experts in Alaska willing to give advice.

"Because many established Alaskans regard pipeliners as carpetbaggers," the magazine said, "*Money* had some trouble finding financial advisers willing to aid Craft.

"The National Bank of Alaska, the state's largest bank, declined to help. Senior Vice President E.B. Erskine said the bank did not wish to publicize high-salaried pipeline workers who are causing problems for Alaska, hurting the economy."

The banker was referring to the resentment felt by thousands of Alaskans whose wages weren't anywhere close to those paid on the pipeline.

After *Money's* editors asked around, John Caven of the Alaska National Bank in Fairbanks, Duff Kennedy of a Seattle investment firm, and CPA Frank Danner of Anchorage agreed to meet with Craft and his girlfriend in his Fairbanks mobile home.

Craft's first question was one repeated by tens of thousands of other pipeline workers: "What can I do right now so I'm not eaten alive by taxes?"

There was no good answer. Craft, like many new arrivals in the upper-income brackets, felt the pinch of a 40 percent tax bite. In reviewing Craft's investment ideas, the advisers agreed that running a business nine months a year would not work because someone had to mind the store year-round.

"Everyone in Alaska has the idea that they'd love to live here nine months and be somewhere else three months," Danner, the Anchorage CPA, told Craft. "I've been here thirteen years, and I certainly would love to spend three months of the year in Hawaii."

The advisers suggested that he put some of his money into stocks, think about business opportunities, speculate in Fairbanks real estate, and start saving about $300 of his weekly net pay of $700.

The session with the financial experts showed that there was no simple way to deal with pipeline prosperity, which was the overpowering force that attracted seventy thousand workers from every state in the union, Canada, England, and various other nations.

"Money's not just an important thing," said a former magazine advertising saleswoman from Anchorage employed on the pipeline, sounding like Vince Lombardi. "It's the only thing."

Brooks Adamson, an ironworker from Montana, said he came to Alaska because he wanted his own "shot at the moon."

"I think the things that have become most important to me are the money and the people I've met," Adamson said. "I've met miners, thieves, bankers. Everybody from college grads to ex-cons are working the pipeline."

Some of the aforementioned would be separated from their money before they got far from the job site by the three most prevalent vices—gambling, drugs, and alcohol. All three were forbidden, but as long as those partaking didn't cause trouble, Alyeska didn't often employ what pipeline project manager Frank Moolin, Jr., called the "overzealous Carrie Nation approach."

Everybody had big plans for their money, but many workers found ways to spend every dime before living their dreams. Then they'd go north and do it over again.

Fairbanksan Ed McGrath, who finished his stint on the pipeline owing others a large amount of money, said the dollars had a way of disappearing fast.

Some workers "have set themselves up for life, some have had a good time, and almost all have been, for thirty minutes at least, the richest man at the bar," McGrath wrote in his book, *Inside the Alaska Pipeline*.

Discriminating pipeliners could order a 10K gold, personalized pipeline ring from an ad that appeared in the *Fairbanks Daily News-Miner.*

Teamster Craig Paradise of East Hartford, Connecticut, knew a lot of people who "ended up working for nothing because they spent their money on drugs or lost it playing cards." A truck driver on the pipeline for two and a half years, he found it was important to stay focused on a goal.

"I'm not out to make a million," said Paradise, who was in his mid-twenties at the time. "I'm saving for a house back home. One more season, and I may have it made." By sending his checks home to his sister, he put away enough to buy a house when he returned to Connecticut.

The financial visions of pipeline workers were as varied as the workers themselves. Thousands wanted to buy land and pay off mortgages, earn money for college tuition, or wipe out mountains of debt. For the first time ever, some people got the chance to stop living payday to payday.

Others bought farms or new cars, started restaurants, and invested in everything from IBM to apple orchards. Attitudes about money changed.

Waitresses in Anchorage and Fairbanks got used to seeing workers dig out $100 bills to pay for a $5 meal.

"Sometimes money burns a hole in your pocket so bad you just have to spend it on something," said Stephanie Leman, a twenty-four-year-old Teamster with land investments in Anchorage. "I used to think a $10 bill was big. Now I pull out $100s and don't think anything about it."

Jody McClarrinon, an electrician apprentice at Valdez, was saving money to study belly dancing in Hawaii. Laborer Frank

Howard bought a two-man submarine that could dive to thirteen hundred feet and was equipped for underwater photography. Texan Tom Hudson, a warehouse supervisor at Old Man Camp, spent $1,000 on a pair of silver-inlaid cowboy boots. Sandy Beaubien, twenty-four, a welders' helper at Happy Valley, bought four acres in Anchorage. Joe Gurschier, a pipefitter at Prudhoe Bay, saved more than $30,000 and planned a four-month vacation in Europe.

Gary Roberts, twenty-two, a welder from Texas, bought a $47,000 Houston home and an $11,000 Thunderbird. "I'll be back for the gas line," he said. Barbara Simpson, a twenty-three-year-old from Arkansas, saved $20,000 and leased fifty-two acres in her home state.

S.A. Steward paid off a $60,000 debt from farming losses in Arkansas. Michael Wardell, a welder at Glennallen, lost $10,000 in the commodities market in 1974, which he said was like playing poker, only riskier.

Welder Harold Rush of North Zulch, Texas was making $1,560 a week, which shrank to $860 after taxes. He kept money for snuff and sent the rest back to North Zulch.

While Rush spoke with a reporter from *National Geographic* in a pipeline camp, a poker game took its course in the adjoining room, with a pot full of $20, $50, and $100 bills. Nodding to the poker game, Rush said: "That's what you might call 'easy go.' But I'll tell you, in this business there ain't such a thing as 'easy come.'"

Ralph Pike, superintendent of the Fairbanks Rescue Mission, was familiar with the workers to whom pipeline pay was "easy go."

"A lot of people don't know how to handle the big money," he said. "Maybe because they make so much they don't put value on it. They'll come to town, either get drunk, get hit over the head and get it stolen, or spend it all uptown some way. We've had them come in here and say they've lost anywhere from $1,300 up to a couple thousand dollars in three or four days time. Of course they've all got a story how they lost it. Then it's a matter of trying to get back out on the pipeline again.

"On the other hand some have appreciated us enough that once they've gotten out, they've come back and made a fairly large donation to the mission," he said.

Gordon Fowler, who had saved most of his earnings as a Teamster warehouseman and bus driver, described two extreme attitudes about pipeline pay: "There were those who lost all sense of value for money. An easy-come, easy-go way of life. They were blowing it as fast as they were making it. It seems as though they thought they'd always be making that kind of money, that there would never be an end to it.

"Then I saw others to whom the money came to mean everything. They hoarded it. It pained them to spend a dime. They couldn't even enjoy R&R," he said in 1977. "In a sense we all got spoiled by the pipeline wages, and it will take a certain adjustment on everyone's part to go back to normal life."

R&R

The majority of workers left camp every nine weeks for "Rest and Relaxation," a term borrowed from the military, with the biggest paychecks of their lives. In some quarters, R&R was known as "I&I," meaning "Intoxication and Intercourse."

It wasn't unusual for workers to hit Fairbanks, Anchorage, or points south with $5,000 to $10,000 in cash. Some workers stayed in Alaska for R&R, but tens of thousands didn't simply because they could afford to go anywhere they pleased— and did.

One worker recalls being offered the chance to go to Rio de Janeiro for a long weekend. The single men and women often flew nonstop to Hawaii from Anchorage. "I just live out of my pocket. I've got a passport and for R&R, I just take off for Europe or Hawaii or home," Rick Lee of South Carolina, a safety coordinator at Prospect Creek, told a reporter in late 1974.

"I don't much like the life, and I hate the work, but I love the money," said Scott Harter, twenty-four, a laborer at Chandalar. Interviewed in September 1976, he already had winter travel plans. "I'm staying here until December 1, when I'll be on a 9:30 a.m. plane to Honolulu. After that I've got reservations straight through to Costa Rica."

Many pipeliners flew the Alaska Airlines "Pipeline Express," a combination flight operated with Braniff Airlines

A Catskinner shows off his civic pride.

that offered one-plane service from Fairbanks to Dallas and Houston, the longest direct flight in the United States. In its bid for the pipeline travel dollar, Pan American urged workers to "Go from the Alaska Pipeline to the Banzai Pipeline" on its weekly $475 tours to Waikiki.

The Alaska National Bank of the North offered the "Pipeliner Account," through which money could be automatically deposited and bank drafts could be sent to families in the Lower 48.

The owner of a Fairbanks pawn shop on Second Avenue, Lazar "Larry" Dworkin, said that checks going through his establishment were "five to three drawn on Outside banks," with most in Texas, Oklahoma, or Arkansas.

He said the money earned on the pipeline was "dissipated at an unprecedented rate," but one big change in the pawn shop business was that those who came to pawn articles were more likely to use nugget jewelry as collateral instead of Timex watches.

Talk of money and how to spend it were constant diversions on the pipeline. The impulse was reflected in Fairbanks and Anchorage, where expensive watches, gold nugget jewelry, electronic gadgets, and other luxuries were much in demand.

"Whenever I come home on R&R, I spend a lot of money on something—anything," a pipefitter told a reporter in 1976. "I've bought more stuff I don't want and don't need just because I can."

Tim Normington, a job seeker who kept detailed notes about conversations with those who waited in line at the Operating Engineers hall, wrote that the pipeline meant money to every man in the hall.

"The men at the hall thread sums of money through their conversations with a sense of wonderment," Normington said. "Two fellows argued for thirty minutes about whether one day of work on the pipeline equals five days or three days of pay back home. For the majority of men the abstract sums of money represent a stake in life—college tuition, land, a house, a small business."

Sandra Foster of Anchorage, whose husband, Pat, was a carpenter in Valdez, described the change that came with pipeline wages: "At first you buy things—things you've always wanted." They bought a four-wheel-drive Chevy truck for $6,300, a $3,000 camper, and long vacations Outside.

"Then," Sandra said, "you realize that the pipeline job isn't going to last forever—only maybe to the middle of 1977. And you need to have something after the boom is over."

Who's Going to Run the Town?

The boom times did not bring the same benefits to everyone in Alaska. Pipeline workers and many businesses prospered, but wages in cities and villages were a fraction of what they were on the pipeline.

Janitors and dishwashers could make $3,000 a month on the pipeline, five times as much as they made in Anchorage or Fairbanks. It was hard to find and keep employees at the lower rates.

Police officers and State Troopers resigned to become pipeline security guards, jobs in which they earned as much in a week as they used to earn in a month. Graduate students quit school to become laborers. Auto mechanics, plumbers, janitors, and others joined the exodus.

The influx of visitors and new residents during the pipeline years created a statewide building boom. Here, new construction is underway in Anchorage at the Captain Cook Hotel.

The turnover rates at many jobs in Fairbanks were probably the highest in the country in 1975. The post office hired 150 people for 210 jobs that year. In six months, the University of Alaska Fairbanks hired 79 people for 77 maintenance jobs. Yellow Cab had a turnover of 800 percent and the Pastime Restaurant, which employed 20 people, had a turnover of more than 1000 percent.

A Fairbanks banker joked that he was thinking of offering diamond seniority pins to tellers who stayed on the job for more than thirty days. The teller turnover exceeded 100 percent at the banks and there was a real fear that when a teller quit, it would be impossible to find a replacement because the job paid only $575 a month, which was not enough to live on. The minimum wage was $2.60 per hour, which translated into $450 a month.

Adding to the stress of these low-paying retail jobs was the sharp increase in the number of customers and the long lines at every store. It wasn't until the pipeline was almost over that the retail and service sector in Fairbanks caught up with the growth in consumer demand.

It wasn't unusual to spend an hour in line at the bank. On Friday afternoons, when everyone wanted their money, two hours was not exceptional. Waiting in line became a way of life at the post office, the doctor's office, the Division of Motor Vehicles, the grocery store, the hardware store, and even at the Fairbanks McDonald's, which was said to be the No. 2 store in the world for volume, trailing only the outlet in Stockholm.

At the Airport Way Safeway in Fairbanks, in one month in 1975, ninety people were hired for the seventy-person work force. The city-owned power plant, staffed with thirty-four people, had forty-eight terminations and new hires in one year.

The Fairbanks dealer of Caterpillar equipment saw a forty-fold increase in the sale of spare parts because of pipeline-related orders. Alyeska and its contractors often bought in bulk at grocery and hardware stores, buying out entire shipments of products when they arrived at Fairbanks stores. There were spot shortages of water softener salt, batteries, four-wheel-drive vehicles, ladders, and other items bought for the pipeline.

Les Dodson, a vice president of Aurora Motors, told a reporter that Alyeska bought all seventeen of the four-wheel-drive trucks he had on the lot. "If I could get shipment of all the models I've ordered, especially in trucks, our business

Charlie Backus/Alaska State Library

Three Irish men outside the Laborers Union Hall in Fairbanks, May, 1975.

would be out of this world," he said.

It was the same with other enterprises. Realtor Vic Hart took a man to see a $60,000 house, and he bought it without going inside. Morrie Jones, owner of Morrie's Body Shop, said he was backlogged with work for three months.

"I put a sign on the door in January which said, 'Closed. Catching up on Backlog,' and I didn't take it down until May," he said.

The feverish pace of business led to many teenagers getting jobs that normally would have gone to adults. High school students in Fairbanks had split shifts, with two separate high schools in one building, one that met in the morning and one in the afternoon. About half the students worked and some dropped out of school to work on the pipeline.

"See all the shiny new cars in the parking lot?" a high school principal asked a visitor. "That is the student parking lot. Now, if you go out the back door when you leave, you will see a parking lot full of old Volkswagens. That is the teachers' parking lot."

The high turnover meant that many people with little experience were being thrust onto the front lines. One visiting reporter said that the "don't-give-a-damn" attitude infected waiters and other service jobs, turning "a casual meal into an ulcer-producing bout with inefficiency."

Trish Baldwin, who had worked as a waitress in a union job, said no one wanted to work in town when you could earn more than double the money with half the hassle on the pipeline.

She said that businessmen could keep good help if they raised the pay of their employees. These debates about what "pipeline wages" were doing to Alaska could spark heated arguments about rates of pay and hours worked. After the *Fairbanks Daily News-Miner* editorialized that "extremely high wages" paid to pipeliners were pushing up the cost of living and "sabotaging Alaska's economy," an Alyeska official wrote a letter to employees saying, "We know, of course, that this is not true.

"If pipeline workers bring home larger paychecks than others," wrote the official, "it is perhaps because they work much longer hours than others."

SECTION IV

POWER AND VICE

The Pioneer

ALL-ALASKA WEEKLY

Vol. 6, No. 4   All-Alaska Weekly, July 25, 1975   Page 1

50¢

At Peak of Boom...

Fairbanks Becomes Wild City, Prostitution Up 5000 Percent

Chamber Ponders Damage to City's Image by Articles 'Outside'

Parade To Climax Celebration

This weekend is the big one for Fairbanks, when the week-long Golden Days Celebration climaxes with a parade that promises to be the biggest and most extravagant in the city's history.

City Council Ponders Clean-up of Second Avenue

By Tom Snapp
Editor

"I knew we were going to get some action when hookers started stopping men walking with their wives and saying, 'Hey honey, why don't you get rid of that old bag and come have some fun with me?"

— Fairbanks City Police Chief Richard Wolfe, on prostitution along Second Avenue

LOOKING FOR LIVE ONES

The world's oldest profession had a long history in the boom towns of Alaska.

As far back as 1909, a federal investigator reported: "The general feeling toward prostitution in Fairbanks is extremely friendly, to such an extent that the majority of the residents know the prostitutes by their first and second names."

For decades, the fenced Fourth Avenue "line" of houses in Fairbanks was home to prostitutes who operated with full knowledge of the community. Prostitution was illegal, but the line operated under rules set by the police department. Each house had to have a rear light on at all times, the women had to take blood tests, stay out of bars and hotels, and not be drunk during "working hours," from 9:00 p.m. to 6:00 a.m.

In addition, to be allowed to work on the line the women had to report to the police every month and to be fined $50 every month for vagrancy before going back to work.

In the 1950s, Roy Webb of the U.S. Justice Department toured the territory's major cities and was shocked to find gambling and prostitution booming outside the city limits of Fairbanks and Anchorage.

"I found the areas surrounding Anchorage to be among the most wicked I have ever visited in all my life," Webb told his superiors.

Like it or not, and there have always been Alaskans with views identical to Webb's, prostitution was part of the gold rush, the military boom after World War II, and the pipeline boom.

"Anchorage has more whores than it knows what to do with—at least eighty-five new faces on the street last summer,"

Michael Rogers, a writer for *Rolling Stone*, said in 1975. "And even the fanciest hotel in town (owned by once governor, once secretary of the Interior Walter Hickel) has hot and cold running prostitutes jumping out of the elevators."

The connection between pipeline boom and prostitution became a fixture in news reports about Alaska in newspapers, on network TV, and in magazines from *Buff Swinger* and *Playboy* to *Time*. In part this was because access to the pipeline camps was limited, which forced reporters to look for something else to write about. The increase in prostitution was as easy to see as the outline of Mount McKinley on a clear day.

"In cafes in downtown Fairbanks, pimps in broad-brimmed leather hats lounge around the bandstand and keep close watch on their girls, up from Seattle and San Francisco for $100," *Time* wrote.

It was this atmosphere that prompted Tom Snapp, editor of the *All-Alaska Weekly*, to write a sensational headline that was quoted in the *Los Angeles Times* and around the world: "Fairbanks Becomes Wild City, Prostitution Up 5000 Percent," Snapp wrote.

Though the headline probably overstated the increase by a few thousand percent, there was no denying that the jump was enormous. The women who looked for business in the summers on Second Avenue wearing platform shoes and carrying large handbags were mostly young and mostly new arrivals from big cities, who were not afraid to advertise on the streets.

An unnamed pimp in Fairbanks told the *Los Angeles Times* that it all stemmed from the money. "I gathered my little flock together, rented us a Winnebago and we headed up the Alcan to the promised land. This is like taking candy from a baby."

In an earlier time in Fairbanks, said Lazar "Larry" Dworkin, the crusty proprietor of the "All American Smoke Shop," prostitution "could have been easily categorized as a 'light industry.'"

But the pipeline invasion demolished the reputation of Second Avenue, and the hookers "engaged in more burglary, rollings, and larceny than they did in being faithful to their ancient profession," he said.

"I think Fairbanksans object to having the whole town represented as 'Sin City,'" said journalist Jane Pender. "But mostly I think they're irritated and annoyed because the present crop of working girls on Second Street is so bold."

Former Fairbanks policeman Steve Porten, who was a law clerk in an office on Second Avenue, said the relatively few pre-pipeline prostitutes kept a low profile. "I'd have lunch with them or a cup of coffee and shoot the breeze with them, it was a fairly loose thing. I knew that they were prostitutes and they knew I was a cop. And they knew that I'd try to bust them if the occasion arose," he said.

With the influx of construction workers and the increased prostitution came more complaints about stolen wallets and related crimes. The ladies of the night rented rooms that were close to the action on Second Avenue. One Fairbanks prostitute claimed to be making $500 a night on two to three tricks, taking them to her hotel room after soliciting outside downtown bars.

"The gals now are newer to the state, and they're only going to stay to make the big money," Porten said in a 1976 interview. "And they don't care that much about establishing a clientele, so to speak, so they're just going to rip people off.

"Now, every ten or fifteen feet you're getting collared by some gal who wants to go on a date with you, a so-called date for fifty bucks. It's just a lot more blatant now than it was a few years ago."

A *Fairbanks Daily News-Miner* reporter was solicited four times in a short stretch on Second Avenue. Like taxis lining up at a cab stand, some of the prostitutes took turns in holding positions on prominent street corners. When one hooked a customer, another would move into place.

One undercover police officer sat down in the Pastime Cafe in downtown Fairbanks and ordered a hamburger. The woman sitting beside him mentioned something about slow service and added "After the hamburger, I'll be dessert." The woman had been in Fairbanks all of five days before she was arrested. "We've always had hookers and drunks down on Second Avenue," Fairbanks police chief Richard Wolfe said. "But the influx of people from the pipeline just caused everything to

Paul Helmar/Alaska State Library

Fairbanks city police escort a woman into their patrol car on a stretch of Two Street near Tommy's Elbow Room.

explode. Nobody knew what the impact was going to be and all of us just sat here with our heads on our shoulders until it got unreasonable.

"It was pretty hard to get anybody excited about prostitution and gambling. Many of our business leaders became business leaders because of the money they made back in the old days from running whorehouses and gambling joints. Everybody knows that," he said.

Opinions changed as the crime became more blatant. In 1974, the Fairbanks police department estimated that twelve prostitutes worked in Fairbanks, charging about $100 an hour. In 1975, the estimates ranged from forty to one-hundred-fifty at the height of the hooker invasion.

The *All-Alaska Weekly* printed an editorial headlined "A Message to the Hookers," in which editor Tom Snapp warned the prostitutes that people in Fairbanks soon would revolt at the brash tactics of prostitutes, who had started accosting pedestrians and taken to flagging down passing cars.

"There is even the possibility that civic-minded citizens may be recruited to make arrests," Snapp said. "In your frenzy for competing for the 'live ones,' you are becoming a spectacle and are threatening to endanger your profession altogether in Fairbanks."

One Fairbanks journalist speculated that the increasing opposition to the wide-open prostitution downtown may have had something to do with the construction of three new malls that would nearly kill off the downtown core area or it may have been linked to people getting "tired of being made to look like ludicrous rustics in the national press."

A thorough *Minneapolis Tribune* report on the pipeline included the observation from a hooker named "Dalene" that there were so many prostitutes that the price per trick was the only thing that had declined during the boom. She said the price in Fairbanks ranged from $20 to $35 and "for good-looking broads, too."

The incessant media focus on prostitution and related problems led to much discussion at Chamber of Commerce meetings over the town's image.

"They never mention the positive aspects of the pipeline," Chamber Manager Wally Baer complained. "The new housing, new industrial development, that unemployment has been significantly lowered."

Frustrated at the image Fairbanks was getting in the press and the complaints from citizens, the city belatedly took action in the spring of 1976.

It came in the form of an ordinance to ban "loitering for the purposes of prostitution," which carried a mandatory sixty-day jail term, the first crime on the city books with a mandatory sentence. The measure increased the maximum fine from $600 to $1,000 and the maximum sentence from sixty days to a year. The typical jail sentence had been ten days.

There had been discussion of giving prostitutes the choice of sixty days or getting out of town, but the city attorney said that was unconstitutional.

The ordinance didn't end the street spectacle, but some of the prostitutes headed to the state's largest city. Anchorage Mayor George Sullivan said that within two days of the approval of the Fairbanks ordinance, the ranks of Anchorage prostitutes increased by about twenty-five.

As to the suggestion that the loitering ordinance was

unconstitutional, Councilman Dick Gruel suggested that the violation be renamed "Moseying for the Purposes of Prostitution."

The weather and the change of seasons also served to encourage some hookers to mosey out of town, as many stayed only as long as the warm summer weather, which wasn't long. On the first weekend in September 1976, "thirty prostitutes boarded the night flight on a one-way ticket to Seattle," according to one report.

It was a short-lived phenomenon, but it became part of the pipeline legend. Cartoonist Milt Caniff dealt with the subject in his "Steve Canyon" comic strip from February through April 1975. The comic featured the exploits of "Pipeline Polly," a long-haired beauty in skin-tight long johns who toured the route of Alaska pipeline camps on her yellow snowmachine selling "encyclopedias."

The cost of each encyclopedia, Polly said, depended on "how much the customer likes to read."

A story made the rounds about a woman who flew to Deadhorse on the North Slope, and in four days returned to Fairbanks with $5,000 and was treated at the hospital for exhaustion. There was also a tale about a busload of fifteen women that was turned back after crossing the Yukon River, headed for pipeline camps.

There were isolated incidents of prostitutes operating rooms of ill repute in pipeline camps. Capt. Tom Anderson, commander of the Alaska State Troopers criminal investigation bureau, described how one scheme worked.

"ID badges from occupants of the camps are rented to pimps, who give the badges to prostitutes so they can get into camp," said Anderson. "The prostitutes then go to their tricks' rooms. They choose heavy traffic times so that there are so many people coming and going the security officers don't notice them."

On the Richardson Highway, one lodge near a pipeline camp was turned into a house of prostitution by a Los Angeles man. The same thing was tried in a house rented in the village of McCarthy, 175 miles from Valdez, where a husband and wife set up shop before being run out of town.

Two of the prostitutes had been left behind without any

money, however, and McCarthy residents chipped in to charter a flight to Glennallen for the two frightened young women, also giving them each $10 in spending money.

One Nevada entrepreneur who boasted he was going to make $1 million traveled north with four women he recruited in Las Vegas. He set up his motor home in the Glennallen campground and put signs advertising rates of $200 an hour at the camp gates.

Rollie Port, an investigator for the Alaska State Troopers, donned a hard-hat and went with another trooper in an Alyeska pickup to the motor home, disguised as pipeline workers.

Port said they were greeted by the Nevada operator with open arms, since they were his first "customers." He did himself in by describing the bank accounts he had opened in Canada and how he hoped to work the entire pipeline. Port was wearing a wire so that two other Troopers in another part of the campground could monitor the conversation.

"The girls were introduced and each in turn solicited an act of prostitution and accepted our money, before we told them they were under arrest," said Port.

The pimp got a stiff sentence, but the four women, who agreed to testify against him, each spent a week in jail.

"They did accomplish their objective, however. All of them got jobs working on the pipeline, but not doing quite what they had originally anticipated," Port said.

Then there was the Paradise Dating Service, which recruited customers from the crowds arriving daily at Fairbanks International Airport.

A man periodically met flights and handed out leaflets offering to make connections with "the prettiest girls in Fairbanks." The leaflets proclaimed that the service was "Operated BY Pipeliners FOR Pipeliners."

Customers who wanted to partake of the dating service, which cost $100 per hour with a one-hour minimum, were asked to called a number and leave an address, time, and "preference of blonde, brunette, or redhead."

It turned out that the service was operated by one man and his wife. She had a selection of wigs. She also had what she claimed was her own lie detector device that she used on

prospective "dates" to identify undercover police.
Various "escort" services opened their doors, offering "dates" for the pipeline crowd.

In a major criminal case, nine people, including two well-connected political figures, were accused by a San Francisco-based Organized Crime Strike Force of conspiring to set up a prostitution and gambling ring in a Valdez bar that was to bring in $1 million in six months. All of the charges were either thrown out or the defendants were acquitted.

The defense argued that the government lured two prominent Alaskans into the plan, apparently a convincing argument for the jury.

For his part, the prosecutor claimed that his case was lost when the trial was moved from San Francisco to Anchorage, to be heard by a home town jury.

In response to the avalanche of publicity about prostitution in Alaska, the attorney general said the influx of hookers was to be expected because of the tens of thousands of construction workers who came to the state. "It is our job to stop these things," said Alaska Attorney General Avrum Gross, "but you cannot delude yourself by expecting the impossible from your police agencies."

In those years before the AIDS epidemic, more people tended to think of the prostitution boom as a natural consequence of the construction boom.

One measure of the live-and-let-live attitude was a 1976 survey in which almost eight out of ten people in Fairbanks said that prostitution and gambling were a matter of "personal morals" and should not be criminal offenses.

A poll conducted by a college sociology class found that Anchorage residents were divided evenly over whether prostitution should be legalized.

Anchorage legislator Bob Bradley introduced a bill to allow a local option of legalizing prostitution, but the bill went nowhere and drew intense opposition. Bradley said the public input was lopsided against it because people in favor of legalizing prostitution were reluctant to say so publicly.

Dealing with the increase in prostitution was also a major issue in Valdez, where there were about forty to fifty prostitutes, and in Anchorage, where there were more than two hundred, according to police and Trooper estimates. On Fourth Avenue in Anchorage, some establishments painted "NO WHORES ALLOWED" on the plate-glass storefronts.

Anchorage Mayor George Sullivan appointed a commission to study the prostitution problem, and it returned with a call to legalize brothels. The commission estimated that 215 prostitutes worked in Anchorage, most of whom worked in massage parlors. Only about twenty-five were much in the public eye on Fourth Avenue, the commission said.

"Some young women have used the profits from prostitution to put themselves through college or support a one-parent family. This being true, and we see no valid reason to dispute it, we must then assume we are dealing with a small minority of Anchorage prostitutes when we view the nightly promenade of Fourth Avenue prostitutes," the commission said.

The group claimed that while brothels should be acceptable for Anchorage, the long-term solution would be to "build better relations between women and men, wherein each partner is encouraged to develop his-her full sexual potential."

The Anchorage prostitutes were on a circuit that included Seattle, Portland, San Francisco, Las Vegas, and Honolulu.

"They generally operate in one area two to three months before moving on to higher-paying and more lenient law enforcement areas," a commission member said.

The Alaska Civil Liberties Union and the National Organization of Women supported decriminalization of prostitution, but objected to the suggestion that the government could regulate the activity. That would be an invasion of privacy, they argued.

The women's rights movement entered into the picture with a decision by Fairbanks District Court Judge Mary Alice Miller to stop sentencing prostitutes because no men were ever arrested for the crime.

The city responded with a short-lived program to go after men, but it ended because of complaints about entrapment.

There were occasional calls, but not many, for Alaska cities to follow the lead of San Francisco, where the district attorney had recently stopped enforcing prostitution laws.

A prostitute in Fairbanks asserted that legalizing the vice would bring in big money and create new opportunities for corruption. Licensing would bring a payoff system, she warned.

Two Anchorage prostitutes told a newspaper reporter they were opposed to the regulation that would come with legalization.

"You'd have to turn a lot more tricks to make any money," said May, twenty-two, a heavyset woman who said she was a prostitute because she would rather make $100 in fifteen minutes than $100 a week.

She said she wanted to make $25,000 to finance the start of a clothing store. Her friend, April, a year older, had a goal of earning $50,000 and going to college. She came to Anchorage from Valdez, where she had operated out of a trailer in 1974.

"I must have made $100,000 that year," she said, "and it's all gone."

THE NORTH STAR
TERMINALS MURDERS

Frank Fitzsimmons, the president of the Teamsters Union, came to Alaska in August 1976 to dedicate a new $6.5 million Teamster office and recreation center in Fairbanks. Getting off the Grumman G-2 jet in Anchorage and walking along a twenty-foot red carpet, he was greeted by Alaska Teamster leader Jesse L. Carr, who in some minds was second only to the governor in the power he wielded in the 49th state. In other minds, he was second to no one.

"You heard we lost two real good guys. We're really upset about it," Carr told Fitzsimmons.

"Yeah, we're really sorry to hear about that. We'll have to take care of that," the union president said.

Almost exactly one year after the mysterious disappearance of Teamster leader Jimmy Hoffa, Alaska had an unsolved mystery of its own with the 1976 killings of two of the leading Fairbanks Teamsters at the North Star Terminals, the main pipeline warehouse in Fairbanks.

The two murder victims were Harry Pettus, 47, the No. 2 Teamster official at the North Star Terminals, and Jack "Red" Martin, 45, the No. 4 union official at the complex.

Investigators theorized that Martin, a convicted felon, was a leader of a drug and fencing operation at the terminals and that he was killed in a power struggle with others involved in the illegal trade.

Pettus, who was regarded as being law-abiding and above reproach, had several disputes with those involved in the illegal trade at the terminals, sources told the newspapers. He kept notes about favoritism in hiring girlfriends and family members, and other practices he disagreed with. "One of these days there's going to be a reckoning," he would say to friends.

After Martin disappeared in July 1976, Pettus began his own investigation, which probably led to his own killing a week later, news reports of the time indicated. "The killers of Martin probably considered Pettus, who had several close friends at the terminals, as too great a threat—he knew too much, had too much power," wrote reporter Tom Snapp in the *All-Alaska Weekly.*

Teamster leader Carr told a reporter that perhaps Pettus had been killed because he "stumbled across something."

Classified ads appeared in Alaska newspapers offering a reward for information about "Harry P.'s confidential report concerning inner workings of NST."

The North Star Terminals had achieved nationwide fame in a much-debated series of *Los Angeles Times* articles headlined "Alaska Today—Runaway Crime and Union Violence."

In one of those stories, the newspaper reported that a score of ex-felons worked at the terminals. The Fairbanks police chief agreed that "some of the toughest hoods" in Fairbanks worked at the terminals.

The *Anchorage Daily News* later called the operation a "tight-fisted Teamster fiefdom" led by Fred Figone, known as "Freddy the Fix."

John Real, the Teamster General Counsel, said that criminal records were no reason to keep a man from working. "Once their debt to society is paid, it would be illegal, if not immoral, to deny these people a chance at rehabilitation," he said.

Most of the workers at the terminals were not ex-cons or thugs, but law-abiding Teamsters who felt unfairly categorized by being lumped in with the ex-cons. According to one estimate by a worker, about 80 percent of the work force at the terminals had college degrees or professional or management experience.

The disappearance of the two men was much on the minds of the workers at the terminals, just as it was on those of the dignitaries from across the country who appeared at the dedication of the new Teamster offices in Fairbanks.

While Fitzsimmons spoke, guards carrying rifles patrolled the roof of the complex. In addition to Fitzsimmons, fifteen of sixteen union vice presidents made the trip. One reporter called their collective appearance on the Alaska tour a "stunning tribute to the rising star of Jesse L. Carr and Alaska Teamsters Local 959."

Teamster boss Jesse Carr was once the most powerful union leader in the state. He died at age 59 of a heart-related condition in 1985.

Local 959 was at the height of its power, controlling an empire that exerted enormous influence over Alaska's economy and its politics. News reports of the time said the union had 23,000 members who ranged from Anchorage policemen to pipeline truck drivers.

The union had two Lear jets, two twin-engine planes, and six pilots. It was building the largest recreation centers in Anchorage and Fairbanks. Its trusts totaled $100 million and were growing at $1 million a week. It owned a hospital in Anchorage, and it had announced plans to build one in Fairbanks for its members.

A detailed examination of the union's power in Alaska by *Anchorage Daily News* reporters Howard Weaver, Bob Porterfield, and Jim Babb received the Pulitzer Prize for Public Service in 1976. The newspaper series characterized the Teamsters as the most powerful special interest group in the state.

A mere decade after the peak of the pipeline, journalists

were writing about the fall of the Teamster empire. The union's power, described in such monolithic terms during the pipeline, was destroyed by many factors, including union excesses and the economic downturn that came with the end of pipeline construction.

The hospital in Anchorage, the jets, and the recreation facilities in Fairbanks and Anchorage had to be sold. The union sought protection under Chapter 11 of the bankruptcy laws and emerged years later as a much smaller and less powerful entity.

If there is no mystery about what befell the union, the same cannot be said of the deaths of Pettus and Martin. Their bodies were found many months apart at two different locations north of Fairbanks, but no one was ever charged.

"One of my biggest regrets is having been too green to do what needed to be done in unraveling the double homicide associated with the Teamsters Union and the warehouse terminal in Fairbanks," said Dan Hickey, the state's chief prosecutor at the time. One big problem, investigators said, was that only a small number of people knew what happened, and they never talked.

The case of the North Star Terminals murders remains open to this day.

"There was a massive lack of security. [Alyeska's] records were so screwed up they had a hard time determining what had been ripped off."

— **Mike Bradner, speaker of the Alaska House of Representatives**

THEFT AND WASTE

"Uncle Al," as Alyeska was known, became a one-stop source of batteries, plywood, insulation, tires, mosquito repellent, first-aid kits, generators, hand tools, food, and other goods for some workers during pipeline construction. It was an inexhaustible horn of plenty.

"Even though pipeline employees were well-paid, many had the feeling that the pipeline company had a surplus of supplies and that a few items wouldn't be missed," wrote Mim Dixon, a Fairbanks social scientist. "When they went home at night, local persons working as craftsmen on the pipeline project at Fort Wainwright lined their pockets, or their trucks, with copper fittings or tools or other commodities with which they had been working. A worker leaving a construction camp with nine weeks' paychecks in his pocket also took with him a pillowcase full of grapefruit from the camp's kitchens."

Dixon saw it as part of a pattern of losing perspective on the value of money. Author Peter Gruenstein said he heard so many workers try to justify stealing that he went out of his way to talk to someone who *didn't* do it.

"I finally found a woman truck driver who didn't, or at least I thought so at first," he wrote. The woman said she thought the theft on the pipeline was really wrong, and she mentioned that the theft would hit everyone in the form of higher prices. When Gruenstein mentioned that people rationalized it by saying that if the items weren't taken they would be destroyed, she said that was exactly right.

"We live in a cabin that a carpenter built with perfectly good plywood that Alyeska was going to destroy. He would just pick up a little piece here and a little piece there. Our whole house was built by Alyeska," she said. "I don't see anything wrong with that, though. They would burn perfectly good plywood all the time."

There were widely varying estimates as to how severe the problems were with theft and waste. A senior Justice Department specialist in organized crime said the rate of thievery "staggers the imagination."

One oil company auditor speculated that losses on the project from theft and fraudulent billings might total $1 billion.

Alyeska officials said the $1 billion figure was "utterly ridiculous." If all of the pipe and all of the camps were stolen, it still wouldn't add up to $1 billion, Alyeska President Ed Patton said.

Alyeska said its records showed that theft was really on the order of $1 million, a figure that, based on the anecdotal evidence of what the workers said, was as speculative as the $1 billion estimate.

The pipeline was built with an emphasis on getting oil to Valdez by the middle of 1977, not on keeping control of tools and other items. Alyeska officials acknowledged this at times.

"Unfortunately, an attitude of 'gold-plating' has evolved," Senior Project Manager Frank Moolin, Jr., said in a late 1974 memo. He said it came about because of environmental and quality control demands, and the weather and terrain. "This attitude manifests itself in 'slugging' the project with people, facilities, equipment, etc.," he said.

This environment of plenty, of "only the best and the most is good enough," often translated into the presence of more workers and more supplies than needed. Patton denied that Alyeska looked the other way when people stole things.

"One of our difficulties is getting people prosecuted in the state of Alaska, where a home town judge does not think it is too bad to steal just a little bit from a large employer," he said.

But law enforcement officials, as well as Alyeska security, said the problems started with Alyeska. Mel Personnett, chief of pipeline security early in the project, said many pipeline officials resisted efforts to crack down on theft. "I've got bumps all over my head from running into brick walls," he said.

Armed guards were posted outside Pump Station 1 on June 20, 1977, when the pipeline was launched into service. At Milepost 0 in Prudhoe Bay, the pipe goes underground briefly and emerges beyond the fence in the background.

"Alyeska is willing to accept a certain level of theft in order to buy labor peace," Attorney General Avrum Gross told the *Los Angeles Times*. "They just want to finish that line. They've stayed about ten miles away from state law enforcement people."

A State Trooper history said the pipeline was "a thief's dream come true" and "equipment was ordered and reordered without question."

The *Los Angeles Times* quoted Patton as saying "there's been more stuff stolen from this project than in the whole history of Alaska." Later Patton backtracked on that statement, saying that "per million dollars of materials put in place, there is probably no more thievery on this project" than on other projects he had been involved in. Given the size of the pipeline project, this didn't really contradict his first statement at all.

"A lot of workers abandon their clothes to use their duffel bags to haul out the stuff . . . like chain saws, binoculars, welding torches, drills, you name it," a former State Trooper working for Alyeska told the *Los Angeles Times*. "I've seen guys dragging duffel bags so heavy they leave a rut in the dirt."

Alyeska officials maintained that on all the really expensive equipment, the theft rate was probably lower than it was elsewhere because it would be difficult to get away with a crane or a front-end loader in remote Alaska, or any other community that's off the road system, like the state capital.

"It's almost like stealing an automobile in Juneau," Patton said. "You can't do anything with it."

Another official stated with pride that "We have found all one hundred and two cranes we have on this project, once again proving that the boom of a crane does not fit in the back of a pickup camper truck."

There was a persistent urban legend about someone who had taken a brand new D-9 Caterpillar bulldozer and stashed it away from the pipeline. "Some speculate that the huge Cat may have been used to excavate its own hiding place, where it will lie until the pipeline's complete and the heat's off," was one columnist's report in the *All-Alaska Weekly* in Fairbanks.

Security guards did report finding a yellow stick hammered in the ground outside one camp. The stick marked the tomb of a $2,000 hydraulic jack.

When columnist and TV commentator Jack Anderson reported that ten miles of the pipeline had been stolen, Alyeska could make a good case that no one had seen hundreds of loads of eighty-foot pipe on Alaska's highways.

Based on the thefts reported to Alyeska and the state authorities, author Robert Douglas Mead said the records "do not support the notion that Alyeska's stores were largely and systematically pillaged in the course of construction."

There still was a question, however, about whether the records provided a truthful record of what was going on, because "Alyeska's requisitioning, purchasing, and inventory systems were several times revised, but remained chaotic to the end. . . ."

Shortly after he became security chief at Alyeska, former Fairbanks City Police Chief Bob Sundberg said he was spending the night at a pipeline camp and could hear people talking loudly into the night at a party next door. The woman was going on about how when she worked at the North Star Terminals, she would tape tools to her body when she went home. She was regaling other partygoers with the drama of one evening when the tools started to slip as she walked out and it was a struggle to get to her car before anything fell out.

"There was a massive lack of security," said Mike Bradner, speaker of Alaska's House of Representatives. "Their records were so screwed up they had a hard time determining what had been ripped off."

Mechanic David McCracken said Alyeska knew that it was being "stolen blind, but did not know how to stop it, did not want to, or declined to pay the price.

"True, they filed charges against a few brazen heist operators, but so many small hand tools and power tools were filched, they were completely unaware of the status of their inventory."

He said that an equipment superintendent at Toolik once spent two weeks trying to straighten out the inventory numbers of 260 space heaters. As part of that effort, McCracken and the superintendent drove two hundred miles to put numbers on a dozen heaters that had been shipped without them.

"I was agog at this frivolous waste of nearly two mandays of time, and the more so when we snooped around another camp and found fifteen heaters which were not on any inventory list," McCracken said. "Any foreman could take a heater, or any other usable property, and if he and his crew were transferred to another camp, usually he took along everything he would need, as well as everything loose that it would be nice to have. If a couple grinders got lost inside of duffel bags, he was completely free to obtain more from the warehouse."

In late 1975, Alyeska security officials met and mapped out plans to combat pilferage. Not long afterwards, Alyeska started inspecting all personal property being sent from the camps. "Before any packages can be mailed from camps, they must first be inspected by a security guard and then wrapped in his presence," said a report in the Alyeska newspaper *Campfollower*.

Empty boxes were being sent to Alyeska camps in the mail system and then returned outside full of stolen tools. All the packages being sent out of the camps had to be sealed in the presence of a guard, which led to a 75 percent drop in the volume of packages.

SECTION V

GROWING PAINS

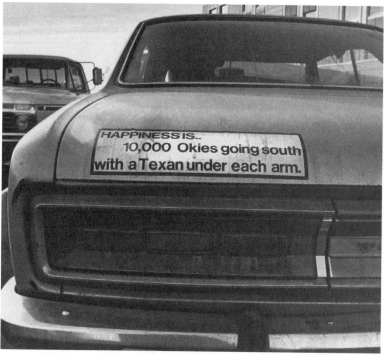

Fairbanks Daily News-Miner

When their hospitality had reached its limits, some Alaskans used bumper stickers to make their point.

"When I came here in 1967 this place was—you know, one of the prettiest places you could ever be. Nobody cared. Nobody was concerned about nothing. No locks, no nothing. It was free. It was total freedom. Now it's—it's completely the opposite, a hundred percent. Everybody's for themselves. Nobody's for nobody else anymore."

— **A Fairbanks man interviewed on the** *NBC Nightly News,* **March 26, 1975**

AN OCCUPIED CITY

Fabian Carey was an outdoorsman and a philosopher in pre-pipeline Fairbanks, a symbol to many in Alaska of the free life and the independent spirit. When Carey ran for local office, he proudly identified himself as a candidate who was "Not Endorsed by the Chamber of Commerce." He won.

He had a gift for storytelling and passions that extended to books, history, trapping, opera, and Dalmatians. He often drove around town with the Dalmatians sitting up in the back seat and looking out at the world.

He was a pilot, a construction worker, a writer, and a Democrat. He was at home in the wilderness, but he was not easy to categorize.

Carey, who came to Alaska in 1937 from Minnesota, had drawn much contentment from living the "boyhood dream of following the far trails and seeking lost horizons."

His way of life was under attack, though, from "Massman," his name for his faceless foe: "Massman the consumer. Massman the accumulator. Massman, who needs a trail bike or a four-wheel-drive vehicle to let gasoline propel him into the wilderness," Carey said.

The men that Carey emulated when he reached Alaska were those who had "turned their back on wages." Their idea of the good life was to earn enough money to afford to go back into the Bush to trap, hunt, and live off the country. Massman was trying to impose a new set of ideas.

"We suffer because Massman can't get it through his head that no one ever owns anything. We only use things for the little time we are here. The only thing a man owns is his life," Carey said.

Carey resisted, but even he couldn't ignore the appeal of a pipeline job.

In 1975 an interviewer took note of the "painful irony" that Carey "has succumbed to the lure of pipeline wages that can climb into four figures a week."

"Some have said it's a contradiction for me to be working on the pipeline construction," Carey said, "since I have been critical of development. But it's too late now. Development is happening, I can't stop it.

"It's pretty hard to stand on principle," he said. "Whether I agree with it or not, as long as Santa Claus is here, easy money is as attractive to me as anybody."

Carey was a member of the Operating Engineers Union Local 302 and was working for the Banister Corporation, a pipeline contractor, when he died of a heart attack at a pipe yard seven miles outside of Fairbanks. He was fifty-eight.

Carey's anguish over the shape of progress and his decision to take part anyway in the greatest invasion of Massman culture that Alaska has ever known reflected the mixed feelings that many Alaskans had in the 1970s. When someone painted the word "Enough" over the population sign on the outskirts of Fairbanks, a lot of people agreed with the sentiment.

The attitudes about newcomers were not lost on newcomers. A respected pollster commented that people moving to the state used to wait about two years before they called themselves Alaskans. During the pipeline, people started calling themselves Alaskans after about two weeks, he said, which was one way to shed the "Outsider" label.

Alaskans, including those who had been on the scene for two weeks, wanted economic prosperity, but they didn't like traffic jams, skyrocketing prices, phones that didn't work, the shortage of housing, and the rest of the fallout from explosive growth.

It didn't hit every town in Alaska with the same intensity. Communities in Alaska's Panhandle might as well have been in another state. In Anchorage, the skyline grew apace, travelers in cowboy boots jammed the airport, and vacant apartments became as hard to find as parking places on Fourth Avenue.

"We've got big city problems we never used to have," Claire Banks of the Anchorage Chamber of Commerce said at the time. "We're choked. Our traffic is now so congested, I can hit any stop sign, and I'll be eight or nine cars back. We don't like that."

Thousands migrated to Anchorage in search of pipeline jobs, but the city was three times the size of Fairbanks and much better able to absorb the sudden impact. Thus an editorial writer in Anchorage, at the height of the pipeline boom, could look at the cramped conditions 360 miles to the north in Fairbanks and say: "It's as if some of the trans-Alaska pipe has fallen squarely on the shoulders of Fairbanks."

Fabian Carey's son, Michael, who later became editorial page editor of the *Anchorage Daily News*, worked on a sociological study in 1974 of how Fairbanks responded to rapid change. He said it was hard to find a "gung-ho developer" that year, and the traditional pipeline booster who wanted unlimited expansion of the economy was rare.

"Even when we go to the Chamber of Commerce, for example, there are a lot of mixed feelings, and people are hesitant," Carey said in 1974. "They say they are concerned about how to keep the old Alaska, or what they regard as traditional Alaska, and at the same time have progress."

Over and over again people said, "I never locked my door in the old days." They were talking about what Fairbanks was like a decade or more before pipeline construction.

Fairbanks Borough Mayor John Carlson was among those who hadn't locked up, but now decided that he had better do so. "Trouble is, I found I'd lost the key. So now I have to take the lock out and get a new key made," said Carlson, who had some misgivings about the pipeline but steadfastly believed "this is an exciting time to be here."

Before the population influx and the traffic jams and the lines that extended out into the street from the counter at Samson Hardware, it was easy to find people who were 100 percent pro-pipeline in Fairbanks. Later on, when it took twenty minutes to drive across town, attitudes were a little different. Residents saw increased pressure on hunting and fishing, and they missed the days when you could walk downtown and say hello to everyone in the small town hard by the banks of the Chena.

Psychologist Jack McCombs said that at first everyone thought Alyeska would be the panacea to all social ills. He called that "the big sugar daddy phase."

"Now there's an overreaction the other way, a growing antagonism," he said.

Elstun Lauesen, a researcher who worked with Carey, argued that people had always come to Alaska as opportunists, but "they stayed because they found an asylum from the madness of America, of Outside."

"There are people who come in now who are not even interested in the community, not interested even in being here except for one thing, take the money and run. And that is bound to irritate the people who respect the traditional community," Lauesen said in 1974.

Larry Carpenter, a former newsman and radio talk show host who'd been hired by Alyeska to handle community relations in Fairbanks, took note of the contradictory feelings people had about the pipeline in a memo to one of his bosses in February 1975. "The attitude of Fairbanks residents toward the project can be termed a paradox. On the one hand the Fairbanksan wants the project because of what it means to the economy of his community; and, on the other hand, he does not want the project because of what it does to his lifestyle.

"For the most part he considers it a necessary evil, which he doesn't mind 'knocking' because he knows it won't suddenly go away and take its dollars with it," Carpenter said.

According to Carpenter, before the pipeline, "poverty in Fairbanks rivaled that of Appalachia," and the economy was on the critical list. The military was cutting back, the state was nearly bankrupt, and the community's largest private employer, Wien Air Alaska, had moved its headquarters to Anchorage. The city was still recovering from a flood that devastated the town in 1967.

In that climate, most of the town's residents cheered when

in 1973, Vice President Spiro Agnew cast the deciding vote in the U.S. Senate to limit court-imposed delays of the pipeline. The boom that followed was the biggest in the town's history, far bigger than that triggered by Felix Pedro's gold discovery in 1902, which led to the founding of Fairbanks. Businesses piled up record profits, and the income of the average resident saw an equally dramatic increase. A 1976 survey found that people enjoyed making more money, but 56 percent said Fairbanks had changed for the worse.

Carpenter said that all of the most critical issues grouped under the heading "pipeline impact" had already been severe in Fairbanks. "Noble Street was bumper-to-bumper during rush hours as long ago as 1970," he said. "The problem wasn't that Fairbanks had too many cars; the problem was that Fairbanks had too few north-south arteries."

Housing was exorbitantly priced and the phone system was broken. "The straw had been dropped on that camel's back," Carpenter said. "The appearance of our project in Fairbanks caused a bale of straw to be dropped on an already broken back."

The people in business, who remembered how economically depressed everything had been, did not see the pipeline boom in such bleak terms.

As former Fairbanks mayor and then Lieutenant Governor H.A. "Red" Boucher put it during a 1974 speech to the Chamber of Commerce: "All of a sudden the dam has broken, the cash registers are ringing, and damn, it's got to feel good."

With few exceptions, most of the town's leaders agreed with the Alyeska representatives, who, in the words of one local activist, advised the town to "sit back and enjoy" pipeline prosperity.

Ernie Carter, who ran a camera and gift store on Second Avenue, said Fairbanks had experienced great economic hardship in the years just after the Prudhoe Bay discovery in 1968.

"Many merchants went in debt to expand their establishments and lay in bigger inventories. Then the pipeline was blocked, oil companies shut down everything and the town went dead. People got scared and quit buying and some even quit paying their bills," he said.

Charlie Backus/Alaska State Library

From Eskimo dolls to pipeline paraphernalia, Fairbanks's airport gift shop was ready for souvenir seekers in April 1975.

People who remembered how close many in the business community had come to financial ruin were willing to put up with the problems of growth. They argued that it was vital if the town was going to survive.

"I think people are willing to accept some degradation of the aesthetics of life in order to keep the economy strong," said Wally Baer, the director of the Fairbanks Chamber of Commerce. "But we're working very hard to keep the lifestyle which brought us here."

The tug-of-war over the wisdom and pace of development was a key factor in the election of Jay Hammond to the governor's office in 1974. After a campaign in which he stressed the need to prevent "malignant development," Hammond squeaked into office by 287 votes over former Governor Bill Egan.

His opponents branded him a "no-growther," while his supporters preferred "slow-growther," but in any event his victory symbolized the misgivings that some Alaskans had about the pace of progress.

A typical Hammond campaign statement: "People come to Alaska to escape many of the things Outside that have degenerated life in their view, and then come up there, and almost in hysterical desperation try to conform as quickly as possible to precisely what they came from."

Hammond's opponents included Jesse Carr, who claimed the governor had filled his administration with people who opposed all development. "They might as well have sent a cat to judge a dog show," the Teamster boss said. Carr often referred to Hammond as a "son of a bitch." Hammond's response to that characterization was to point out that while both he and Carr had served in the Marine Corps, he hadn't known they had anything else in common.

Biologist David Klein said that although Alaskans debated development and economic prosperity all the time, it wasn't money that kept people in the state over the long run. It was an appreciation for non-monetary values like wilderness and the feeling of being uncrowded.

"Yet, when it comes down to decisions, they take these for granted," Klein said. "They don't feel that they have to protect these other values, that they're going to be there, and you can have your development and you can have your low population density and opportunities for outdoor recreation . . . along with development. This is where the conflict exists, I think."

It was easy to find people who said the pipeline was doing irreparable harm to Alaska. Some of them had been in Alaska fifty years, others only a few.

"Most of the changes are negative, unless you're employed by the pipeline and raking in a lot of money," said Janet Baird, a former school board member in Fairbanks. "I think in the long run what we're facing is a change of values, a way of life we've gotten used to in Alaska. We simply won't have that low-pressure, noncompetitive atmosphere any more."

Tom Snapp, the editor of the *All-Alaska Weekly*, believed the pipeline was destroying the town. "Fairbanks has not just become a slightly worse place—it has become a horrible place to live and work," he said.

The manager of a trailer park in Anchorage said it had done the same thing to that city. "This city's going to go to hell, especially if the gas pipeline comes in after the oil pipeline," he said.

The Reverend Don Hart of St. Matthew's Episcopal Church in Fairbanks said the arrival of so many people with no experience or knowledge about Alaska created some of the conflict. "There's an intimacy that is getting stretched pretty thin. I think that was a mark of northern Alaska," he said.

Even the stridently pro-development editorials of the *Fairbanks Daily News-Miner* occasionally took a more skeptical tone when the pipeline invasion was in full swing. No longer was there a vision of unbridled prosperity for all. In February 1975, the newspaper likened the city residents "to the children of an Indochina hamlet standing on the roadside waving and cheering as the troops drive past after having destroyed their village only hours before.

"The children's lifestyle has just been turned upside down, but they are happy because the soldiers are passing out chocolate bars and giving free rides in their Jeeps while all the time driving on toward something much bigger," the editorial said.

A year later, the *News-Miner* announced that it would no longer be printing its annual "Progress Edition," which had been a tradition for a quarter-century. The idea of the Progress Edition was to attract industry and new development to the Fairbanks area. "We do not need more 'progress' at the moment," the *News-Miner* editorialized.

Joe LaRocca, a journalist hired as head of the borough's Pipeline Impact Information Office, said that the *News-Miner* and the business and political leaders of Fairbanks hadn't paid any attention to detrimental effects that might come with the pipeline.

"There's no question that a massive change of lifestyle is about to descend on us," LaRocca said in the summer of 1974. "That wasn't considered at all initially. No meaningful consideration was given to the deterioration of the Alaskan lifestyle."

Allusions to wartime conditions and an invasion by an occupation army riding in Alyeska Pipeline Service Company's yellow pickups were frequent.

"It's like living in an occupied city with helicopters flying around in formation and truck convoys plugging up the road," one resident said.

Fairbanks Daily News-Miner

The lawn outside Fairbanks's Visitor Information Log Cabin is bordered by two of
the busiest streets in town, but this napper was oblivious to the noise in the spring
of 1975.

Twenty-eight-year-old surveyor Sumner Putnam vented
the frustration many felt when, in the spring of 1974, he used
a stencil to paint "YANKEE GO HOME" on the trunks and
bumpers of cars with out-of-state plates.

Someone suggested the bumper stickers and water soluble
paint would be just as effective in this campaign, and less
offensive in the eyes of the law.

The bumper stickers popular in Fairbanks showed something of a change in attitude over time. First there was "Let
the Bastards Freeze in the Dark," a slap at pipeline opponents
in the Lower 48 who blocked the pipeline, and then "Sierra Go
Home." Environmentalists responded with "Freezing in the
Dark Builds Character."

Within a few years, it was more common to see "Alaska
for Alaskans, Yankee Go Home," "We Don't Give a Damn How
They Do It Outside," "To Hell With the 48" and "Happiness

is 10,000 Okies Going South with a Texan Under Each Arm."

Psychiatrist Harold "Doc" South said there was a competition between the people who had lived in Alaska a while and liked it the way it was, and those who arrived on the leading edge of change.

"The new group—the out group—is competing directly with the in group—the Alaskans. This leads to bitterness and feelings of rejection on the part of both. . . . It looks like there will be a long-term change in lifestyle here, and perhaps people should not think of this as a temporary thing."

The Reverend Gene Strattmeyer, pastor of the First Presbyterian Church in Fairbanks, said the pipeline brought problems, but it also led to improved services.

"We're slowly losing the frontier type of village," he said. "I'm sure the people who were here ten years before I was would have said the same thing. I just looked at a picture of what Fairbanks was like in 1904. I think if anyone was here in 1904–05, we've come a long way. I still think that in spite of the change, Fairbanks maintains a special flavor of its own. Maybe other pipelines will sort of annihilate that frontier spirit, that sense of friendliness. I don't think that's been totally wiped out by the pipeline."

Alaska historian Robert DeArmond offered an insightful observation when he said the experience of Alaska in the pipeline boom triggered emotions similar to those from the stampedes that brought an influx of new people a century ago. Even sourdoughs who hadn't been in Alaska ten years were put off by the new arrivals.

"Alaskans who were on the scene at the time of the great Klondike Gold Rush asked, 'Who the hell are this riffraff coming into our country?'"

The same question was asked when the military bases were built, when the Alaska Highway was constructed, and when the pipeline was built. You may hear it the next time Alaska goes boom.

Fairbanks Daily News-Miner

At Valdez, construction on the Marine Terminal was taking shape by May 1976.

> *"Sure there are people here because of big money. But others are starting up businesses here because they love it. They are part of an historic event. This terminal over here will put the pyramids to shame."*
>
> — **John Kelsey, former mayor of Valdez**

IN THE PATH OF THE PIPE

The other boom town in the path of the pipe was Valdez, which, like Fairbanks, was born of the Gold Rush and had seen good times, hard times, and natural calamity. The great Alaska earthquake of 1964 spawned a tsunami that destroyed the waterfront and forced the town to rebuild on safer ground four miles away.

After years of economic struggle and dislocation, the survivors of the Valdez tsunami had a tendency to look upon oil as a pipeline to economic salvation. Writer Calvin Trillin of *The New Yorker* visited Valdez in 1974 and found that townspeople could hardly wait for the boom.

"It has always been taken for granted in Alaska that a town touched by a grand project has found success," Trillin wrote. "Wanting a project like the pipeline is so much a part of the Alaskan culture that some Valdez residents seem to feel mildly guilty at harboring any regrets at all about the end of their quiet winters or the beginning of what will sooner or later be a company town. How could Valdez have accepted the alternative—to be left out of the greatest private construction project in the history of man?"

The answer was, it couldn't.

Spectacular scenery engulfs Valdez, justly described in the tourist brochures as the "Switzerland of Alaska." It is known for fantastic fishing and for wet weather that brings an average of twenty-five feet of snow a year. The town, 115 air miles and 300 road miles from Anchorage, is nestled between the waters of an ice-free fjord and the Chugach Mountains, which climb sharply to the sky.

In 1974, Valdez had two grocery stores, a bank, a clothing store, various offices, hotels, a cafe and restaurant, a post office, a city hall, and schools.

The pipeline transformed the sleepy little village into one of Alaska's most important industrial centers and the only port in America that shipped out oil instead of taking it in.

Across the fjord from Valdez, a place known by some construction workers as "overseas," a work force that peaked at 4,300 built the $1.4 billion Valdez Marine Terminal. Fluor, one of the largest engineering contractors in the United States, had the contract.

The complex included control buildings for the entire pipeline, sewage plants, power plants, metering systems, facilities to treat 400,000 barrels a day of ballast water discharged by the arriving tankers, eighteen colossal storage tanks, and four berths for tankers to load oil at up to 100,000 barrels an hour.

Pipeline construction did more than anything since the earthquake to remake the town. The population skyrocketed from 1,350 in January 1974 to 6,512 by the summer of 1975, and topped out at 8,253 in 1976.

Housing was in short supply and the sewer, water,

telephone, and electric systems were stretched to the limits. Homes that sold for $40,000 one year could bring more than $80,000 the next. Mobile homes popped up everywhere. Itinerant job-seekers lived in cars, trucks, and campers parked here and there along the roads and streets.

One sign of the business boom in Valdez came when the manager of the Club Valdez took out a classified ad in a Seattle newspaper, looking for a tough bouncer.

The ad read: "Tavern Bouncer person wanted. Must be 6' 8" or over, ugly, tough and mean, but diplomatic, pref. 280-300 lbs. The bigger, the better. $50 shift, room and board. Will pay air fare to Valdez, Alaska."

The manager, David Huffman, said he took out the ad because Valdez had "a bunch of gorillas up here that like to fight."

As many as 3,480 workers lived at the Terminal Camp. Had it been a city, the Terminal Camp would have been on the top ten list of Alaska settlements.

"Fairbanks was crowded," pipefitter Potter Wickware said, "but Valdez is like a D.W. Griffiths mob scene."

The president of the Valdez Chamber of Commerce decried reports of rent gouging, blaming those on Anchorage investors who bought up property in Valdez. He added that business was great in Valdez, however, and that the profit potential was possibly unmatched anywhere else in the nation.

Owen Meals, a Valdez pioneer who had arrived as a child in 1903, had provided the land for the city to rebuild on after the earthquake. He told a reporter his biggest thrill was to look across the bay from the city of Valdez with binoculars and watch the terminal take shape on the thousand-acre construction site.

"Everything is either growth or death, you know. There's nowhere you can ever go to avoid change," said the eighty-three-year-old Meals.

John Kelsey, a former mayor who had promoted Valdez as a terminus for the pipeline, said the project was vital to Valdez. "It's bringing us a better way of life, things we never had before: doctors, dentists, a drugstore, a hardware store, radio store, a six-month interim radio station," he said. Kelsey looked ahead to the day when more pipelines would be built from the North Slope and perhaps a refinery in Valdez.

Not everyone was so exuberant, however. John Rogers, a fisherman and a construction foreman, said he would be buying Christmas gifts from stores instead of catalogs, but he had reservations about what was going on.

"We used to live slowly from year to year, getting our bills paid by winter and then relaxing until spring," he said. "You'd make good money, but you didn't have to work as hard to make it. The pace wasn't so frenzied."

New stores and offices opened overnight, selling everything from motorcycles to trailers. In the midst of the pipeline, the high wages in construction jobs were a constant topic of conversation.

"Camp workers often joked about how little they did for their high salaries, and others complained of the high cost of living brought by both these wages and the spiraling labor costs that had spread throughout Valdez," sociologists Michael Baring-Gould and Marsha Bennett wrote in a report on what happened to Valdez.

The median income in Valdez more than doubled in one year, but prosperity came at a price, as it did in everywhere else.

The Valdez phone system, which had twelve telephone circuits and 1,114 telephone lines, shot up to forty-four telephone circuits and 4,262 lines. As in Fairbanks, the phones were constantly busy. "If I start out trying to call Anchorage at 1:00 p.m. I might get through by five o'clock," said Mayor Emil Wegner.

Until a second grocery store opened, some residents shopped in Glennallen, which was more than a hundred miles away. There were spot shortages of everything from toilet paper to milk and soap.

In contrast to Fairbanks, the town retained a more positive attitude about the oil pipeline despite the problems of growth. In part, this was because the population increase was so rapid that the people who had been in Valdez for more than two or three years had become a minority.

"There is but little doubt that the average Valdezian bears a significant commitment to continued growth," the sociologists said.

City lots that had sold for $400 in the late 1960s brought

$4,000 in 1973, $8,000 in 1974, and $10,000 in 1975.

An article in *Playboy* magazine, denounced in Valdez because of the dutifully exaggerated details about drugs, alcohol, and prostitution, quoted a Valdez worker disgusted at the price increases.

"I'll tell you one damn thing. If you pick up something in this town, don't set it back down. Because if you do set it back down even for a minute, it'll be another price when you pick it up again."

The pipeline was finished in 1977, but like the rest of Alaska, Valdez did not return to what it was. The population dropped to 3,000 by 1980, but the town prospered.

The taxes on the oil complex and the jobs created by the terminal operation helped make Valdez one of the wealthiest towns in Alaska.

"I think Fairbanks will grow only as fast as the houses are built. That's why Fairbanks hasn't grown before."

— **Dennis Wise, Fairbanks developer**

SQUEEZING IN

A two-bedroom house a few blocks from downtown Fairbanks was home sweet home for as many as forty-five people in the spring of 1975. The rent for a bunk at the Al Cannon rooming house was modest—$40 a week, one-fifth of what the newest hotel downtown was charging. There were beds in the kitchen, five beds in what used to be the living room, beds in the basement, and beds in the two bedrooms.

It wasn't much different for thousands of people who were living in close quarters in Fairbanks, Anchorage, and Valdez at the height of the pipeline boom.

The beauty of Cannon's rooming house, according to the satisfied tenants, was that the men worked different shifts, so they were never all home at the same time. Of the thousands who moved to Alaska to find jobs on the pipeline, many found such rooming houses were the best and cheapest way to get a roof over their heads.

"I enjoyed the company of some interesting people while 'bunking with five other human sardines' in that ex-living room," said Len Will, a pipeline worker on the North Slope.

Cannon didn't deserve all the grief he was getting, which included national media attention from *60 Minutes* and others. He deserved praise for helping alleviate the critical housing shortage, tenants said.

"Sure they're making money, but so did Ford, Rockefeller, Hunt, Lady Bird, and a lot of other people," Will said. "During the rapid expansion of Fairbanks, some bending of regulations is necessary—within safe and sensible limits."

Robert Truett, another roomer at Cannon's house, said that many times people were weeks behind in paying the rent, but Cannon didn't kick them out. Truett said he found lifelong friends at the house, and the landlord "gave us a fighting chance to get on the Slope."

However, the Fairbanks North Star Borough zoning authorities didn't believe the rooming house was safe and sensible, and they obtained an injunction to shut it down. After that there was no room at that inn.

In another Fairbanks rooming house, a single-family home had been transformed into accommodations for twenty-six job seekers.

Some boarding houses, with rates ranging from $40 to $70 a week, were clean and offered a bunk and perhaps a place to shower. These were billed as "semi-private" accommodations, and the owners occasionally offered "all new box springs and mattresses."

The rooming houses boomed because, as was often reported, "Fairbanks was 103 percent full," based on the housing occupancy rates.

"There are places we normally would move to close up," said state health sanitarian Frank Vonder Haar. "But the real question is, should we? You've got to ask, who are you helping? What do you do, kick someone out in the snow?"

Horace Lewis, a Fairbanks zoning officer, voiced similar sentiments: "It's almost a necessary evil. These people have to stay somewhere."

Adam Small, who had a Fairbanks house with twenty-six bunks, said he was taking in about $4,000 a month, minus $1,600 a month in expenses.

"I have had offers from men to pay me $5 per night to have them just put their sleeping bags on the floor for the night," he said. "However, since I had the word that no more people were to be moved into the house, I had to refuse them."

A log building just south of town became a pipeliners' hotel with tiny sleeping stalls. In the winter, floor space cost $10 a night, plus $5 for pillow and blanket. The rate was $20 to sleep on the old pool table and $10 to sleep beneath it.

Two-room log cabins with two five-gallon water jugs as the only plumbing fetched $500 a month. Families took in friends and converted sheds, basements, and garages to apartments. It wasn't unusual to have ten people living in a two-bedroom apartment. A man with used buses for sale ran an ad saying that the buses could be turned into houses.

More than a hundred people signed up on the waiting list for apartments in the eight-story Northward Building in downtown Fairbanks. When apartments elsewhere opened up, sometimes there were bidding wars among potential tenants. One apartment owner said his vacancy rate was "about four hours." Because of the added wear and tear on the buildings, many landlords would not rent to people with children or pets.

As bad as the housing shortage was in Fairbanks, it was much worse in Valdez, where the population doubled and doubled again. Mobile homes had makeshift tarpaper-and-plywood shacks attached for extra space.

A visitor to Valdez who couldn't find a room in a motel spent her first night bundled up in her car while the wind blew at 75 mph up from the Gulf of Alaska. There were no rooms available even 115 miles away in Glennallen.

She listened to the Valdez City Council discuss the idea of cracking down on transients who were sleeping all over town in their cars.

"I found the discussion strangely personal," she said. "They were talking about me."

One of the people who was living in her car parked her blue two-door sedan outside the Valdez Laundromat. She had friends staying nearby in a camper and had arranged a system so that if she had trouble at night, she'd start honking the horn to alert them.

In December 1974, Kathy Reid and another Valdez teacher rented a two-bedroom apartment for $286 a month. On March 1, the rent went up to $520 a month. A month later, they were told the building had been sold, and the rent was going to be $1,600 a month, plus an additional two new roommates.

"If you aren't into big money, and school teachers and others not employed by the pipeline companies aren't, you can't afford the rents," Reid said. "The whole town is upset, but nothing is being done to combat the problem."

The Valdez city manager agreed and said the situation was atrocious.

"The fact that there isn't any housing available puts the existing housing in extremely high demand and people all over

town are being evicted because they can't pay the high rents,"
said Herb Lehfeldt.
One critic said that with the housing crunch, "Valdez had
become a very carnivorous town." The superintendent of
schools decided to stop interviewing prospective teachers out
of fear they would have no place to live.
The shortage grew so bad that state authorities declared
housing emergencies in Fairbanks, Valdez, and Anchorage. A
lawyer who worked for Ralph Nader said the effort was a joke,
and that the provisions that allowed landlords the ability to
factor in an extra percentage for possible vacancies was like
"sending shipments of oil to the Shah of Iran."
The state set up rent review boards in the three cities to
hear complaints about rent gouging. The boards reviewed the
landlords' financial records and limited profits to 15 percent.
Commerce Commissioner Tony Motley, using an anal-
ogy particularly suited to Alaska, said he wanted the rent board
to "take a rifle approach to aim at the guy who was gouging,
not the shotgun approach to rent control that has been used in
places like New York City."
Over a year, the Fairbanks board heard complaints filed
against 79 landlords by 150 tenants, about rent increases and
evictions. The board settled about half of them, while the
others were withdrawn or settled privately.
For all the uproar over housing, most landlords did not
engage in rent gouging, and many of them had legitimate complaints about tenants that never attracted much notice. For
every story about gouging, there could have been two or three
about tenants destroying property, violating leases, leaving
trash behind, and skipping out on the rent.
One woman who converted her family's Fairbanks split-
level into a duplex said the tenants routinely used up all the
hot water with half-hour showers, allowed the pipes to freeze,
and trashed the apartment. She made a profit of $236 in six
months.
"There were more repairs needed on the house in two
months than the four and a half years we lived in it," she said.
"Are we going to raise the rent? You'd better believe it."
A housing shortage had long been a problem in Fairbanks,

An insulated lean-to expanded the living space for this family at Four Seasons Trailer Court in Anchorage, December 1975.

and there was little new construction from 1969 to 1974 to absorb the thousands of new arrivals. The fixer-upper market was virtually wiped out by a flood that devastated Fairbanks in 1967, and housing was in very short supply during the brief economic boom that followed the discovery of Prudhoe Bay.

It was much worse when the real boom began. Because of high interest rates and the difficulty of obtaining financing, little rental housing was constructed. It was almost impossible to get mortgage money to build apartments.

"This is unbelievable that in this boom year in Fairbanks we are seeing so little in apartment construction," said Realtor Jeff Cook, president of the Fairbanks Board of Realtors.

One of the few people building apartments was Dennis Wise, who was on his way to becoming the largest apartment owner in Fairbanks, having doubled the number of units he owned in each year of the pipeline boom.

With the shortage of housing, lines of people formed at the *Fairbanks Daily News-Miner* to grab the first copies of the afternoon paper as soon as it was printed to see if by some miracle any apartment vacancies had been advertised.

One ad offered a home that was described as the "Buckingham Palace" of Fairbanks for $850 a month, but warned the "meek and the poor need not apply." One critic said the owner should make it an even $1,000 and throw in a chauffeur and a maid.

Other ads advised callers to "make offer" on an apartment or to be prepared for some home improvement work. "Rentee must install plumbing and sewer lines themselves," one ad said.

Apartments could be rented before they were built. A city employee who was remodeling a duplex recalled the woman who walked in and said, "I'll take it." This was despite the knowledge that it wouldn't be ready for several weeks. "That's okay. I'll take it," the woman said.

People also said, "I'll take it," for fictitious apartments. A company started advertising for tenants for a new apartment complex in South Fairbanks. Twenty-six tenants desperate for a place to live put up $500 deposits more than two months before the buildings were to open.

One of those who paid was Gary Stein, a historian at the University of Alaska. Stein drove by what was supposed to be the future home of Fairwood Apartments in South Fairbanks several times in the weeks that followed, but never saw any sign of activity at the building site.

There never was a plan to build and the company didn't own the property. The case went to court and the would-be tenants got their $500 back a year later.

One incident harkened back to the World War II boom in Anchorage, when government officials complained that it was impossible to keep track of the housing supply because houses were being moved so often and there were no forwarding addresses.

Ginger Noteboom bought a one-room $200 cabin off the Steese Highway north of Fairbanks that had to be moved. But when Noteboom and her sister and brother-in-law, Janet and Ray Duncan, arrived to take custody of the cabin, it had disappeared.

It turned up some time later. The cabin had been moved twenty miles, and a young man was living in it who said he had found it alongside a road.

Fairbanks legislator Steve Cowper, in arguing the merits

of legislation to make it easier to get mobile home loans, said, "Up my way we have people renting tent spaces for $150 a month, and that's if you supply your own tent."

Every summer night, a state campground on the Chena River in Fairbanks attracted a full house of people looking for pipeline jobs. On busy days, the rangers turned away a hundred cars or more. One visitor counted sixty-one tents and RVs crowded in addition to those filling all the designated camp sites.

One man who owned an eight-foot cabover camper took out an ad in the "Houses for Rent" category of the classified ads. He wanted $150 a month for the cabover.

The demand for housing led to a big jump in mobile home sales in Fairbanks, Anchorage, and Valdez. Selling for an average of $25,000, about twelve hundred mobile homes were brought into Fairbanks in two years, about half of the new housing units added to the community.

Borough Mayor John Carlson said at one point that if a company could bring four hundred mobile homes into Fairbanks, it would be able to sell them all in one day. The salesmen were exuberant.

"We were like blind poets," said Mike Woldman, the general manager of Columbia Mobile Homes in Fairbanks, describing the nearly $1 million worth of trailers his firm sold in three months in 1975, "selling trailers we had never seen from floor plans we didn't understand." There was a shortage of mobile home parks, though several new ones were built.

In one of the most peculiar responses to the housing crisis, a family of four moved into a parade float built like a giant phone.

"In the ear piece they had shoved a mattress and a kid was sleeping there with two dogs. There was a man, his wife, and two teenage sons," said the owner of the phone, used to advertise the secretarial service Girl Friday.

It was phone sweet home during the pipeline boom.

ROAD WARRIOR

In 1975, Alaska's crowded highways took a beating from pipeline construction, so much so that Alyeska Pipeline Service Company bought radio time to warn motorists to stay off the roads between Fairbanks and the Yukon River.

With more than three hundred pipeline-related trucks and cars on those roads daily, it was dusty, crowded, and dangerous, as Alyeska pointed out in its "Attention Motorists" advertising.

"For your safety, Alyeska Pipeline Service Company urges you to avoid driving the Elliott and Yukon Highways if at all possible," the company asked.

For Fairbanksans who always placed a priority on making the most of summer travel, the request sparked resentment, even though everyone could tell that it was a safety issue.

Fairbanks attorney Lloyd Hoppner responded by taking out radio ads of his own. They sounded almost exactly the same as Alyeska's, but proposed a different solution to the crowding crisis.

The script for Hoppner's ad read: "Attention Alyeska: This summer more than thirty thousand Alaskan taxpayers will be traveling the Elliott and Richardson Highways, highways that have been built and paid for by them. Because of heavy truck traffic, these highways have become extremely dangerous to these Alaskan taxpayers.

"Therefore, for the protection of these Alaskan taxpayers, I urge Alyeska to avoid driving Interior Alaskan highways if at all possible, and if Alyeska must drive these highways, please don't take your half out of the middle. This ad has been paid for by Lloyd Hoppner, who thought you would like to know."

A lot of people in Fairbanks thought the ad was funny, but the *Fairbanks Daily News-Miner* and Alyeska spokesman Larry Carpenter did not. The *News-Miner* said Hoppner "thought his broadcasts were humorous. Perhaps so, to a very limited degree."

Carpenter said in trying to be cute, Hoppner was "halting a safety program."

Hoppner said that Alyeska's attitude in asking the public to stay off public highways was that "Big Oil can do no wrong." With dueling highway ads on the radio, Alyeska decided to halt its radio commercials, but continued its newspaper ads.

"Alyeska should take a few of the millions of dollars they are wasting around here and build their own roads and get off of ours," Hoppner said in 1975. It seemed to him that Alyeska was telling Alaskans to "just crawl in your holes for two years, and let us do our thing."

"All of a sudden, living in Fairbanks isn't fun anymore," Hoppner said. "It's becoming intolerable."

"Dodge City a hundred years ago had nothing on Fairbanks during the peak of pipeline construction."

— **Jay Yakopatz, shift corporal for the Alaska State Troopers in Fairbanks from 1972-76**

FROM SECOND AVENUE TO TWO STREET

A short stretch of Second Avenue, which became "Two Street" to the pipeliners, long had been the commercial center of Fairbanks—the main street of a busy downtown district.

Visitors in from the villages liked to meet outside the Co-Op Drug Store, the most popular rendezvous spot in town. Bankers, trappers, construction workers, and housewives congregated in the three dozen retail shops downtown.

The atmosphere changed in dramatic fashion during the construction of the pipeline, however, and many people began to refer to Second Avenue and environs as a "skid road." Pimps, prostitutes, drug dealers, and gamblers arrived in force, lured by the promise of big money.

Second Avenue was home to restaurants, banks, jewelry stores, a shoe store, a clothing store, a music store, a book store, insurance offices, borough offices, dentists' and lawyers' offices, a TV and radio station, federal offices, the post office, two weekly newspaper offices, a sporting goods store, taxi offices, a barbershop, a movie theater, gift shops, and pawn shops.

But by far, the most numerous enterprises downtown were the bars. In 1975 there were four package liquor stores and eighteen bars on Second Avenue or within a couple of minutes walk. Downtown watering holes between First and Third avenues included: the Chena Bar, Tommy's Elbow Room, the Savoy Bar, the Cottage Bar, the Roustabout, the Riverside Bar, the Persian Room, the Hideaway, the Mecca Bar, the Flame Lounge, the Fairbanks Bar, the French Quarter, Chilkoot Charlie's, the Pastime Bar, the Top of the Pole Cocktail Lounge, the Polaris Lounge, the Northward Cocktail Lounge, the Stampede Saloon, the Gold Rush Saloon, and Ken's Pipeline

Bar. This in a town that also had the Prohibition Liquor Store, located in the Roaring 20s Hotel, the decor of which included what the proprietor said were the original paintings of "Dogs Playing Poker."

There were bars for those who wanted to see "exotic dancers," bars that catered to Native Alaskans, bars that attracted pipeliners, and bars for those who wanted to hear rock 'n' roll or country. Most were in old rundown buildings that contained false fronts, neon lights, and crowds of men looking to blow off steam.

"When I look at Second Avenue, my heart pounds," said German Gerhard Konitzky, the vice president of the company that launched Prinz Brau, the new Alaska beer brewed in Anchorage. "This is really a beer-drinking town. You can feel it."

You could also smell it, see it, and experience it first-hand during the pipeline boom.

The rowdy reputation of Two Street was spread by almost every news report about what happened in Fairbanks during pipeline construction. This caused great dismay in Fairbanks, because it wasn't an accurate generalization about life in the city.

Even Two Street was, in fact, two streets. By day it was frequented by those who wanted to buy film at the Co-Op Drug Store, visit the borough mayor's office in the Lathrop Building, eat lunch at the Star of the North Bakery, or buy a magazine at the Borealis Book & Gift Shop.

At night, it was the gaudy influence of the bar scene that gained the most notoriety during pipeline construction and became the biggest show in town. As things got worse, however, the novelty wore off. Many Fairbanksans who did not want to go to the bars stopped going downtown at all if they could. This attitude nourished the drive to develop three new malls that replaced downtown as the retail center after the pipeline.

At one point, fifty downtown business people signed a petition that said: "We must cleanse the downtown core of the many undesirable prostitutes, drunks, vagrants, sleepers, and loiterers that tend to clutter the streets, particularly during the summer months."

Tommy Paskvan, who operated Tommy's Elbow Room on Second Avenue for more than forty years, described the pipeline boom as being "like a guy going out on a spree having a bottomless barrel of money.

"It was rather prosperous," he said.

Pipeline workers with huge checks would cash their checks and head downtown to begin enjoying "Rest and Relaxation," a misnomer for much of what transpired on Second Avenue.

Drugs, alcohol, and prostitution were the common denominators for thousands who spent time on Second Avenue during those years. It was a latter-day Gomorrah to some, a spectator sport to others.

Jeanne Wilson, an Alaska Airlines employee who was active in the Chamber of Commerce, talked about the time she drove downtown with her daughter one July night at 10:00 p.m.

"We watched one of the prostitutes not only proposition a man, but watched her lift his wallet right out of his pocket," Wilson said.

Pimps circled the block in their Cadillacs, and Alyeska's yellow pickup trucks were much in evidence. Traffic moved along slowly in the congested city center, which only improved with a new expressway finished after the boom was ended. Hookers stopped men on the street, while the Chamber of Commerce had to constantly chase away sleepers from its lawn along the muddy Chena River.

"The Flame is the hardest place to work," a bouncer told a reporter in the summer of 1975. "At the Flame you're talking about a bar that caters to men to show off women, and that means trouble."

A nineteen-year-old dancer who worked at the Flame Lounge and the Bare Affair, a bar just outside the city limits, said the pipeliners were loud and fun to party with. The Flame was a seedy place, she said, but the Bare Affair was trying to appeal to a different crowd.

"The Bare Affair had a rule that you couldn't yell 'Yahoo' or you'd get thrown out. We were trying to be more refined," the dancer said.

City police said that 80 percent of the prostitution arrests took place in or outside the Flame and the French Quarter,

Fairbanks Daily News-Miner

The Flame Lounge gained a reputation as one of the rowdiest of Second Avenue's bars.

which were favorite places for prostitutes and drug dealers. The two bars became a "must see" for every visiting reporter in Fairbanks.

Dancers at the two bars had been interviewed by, among others, reporters for CBS, NBC, the *Seattle Times*, the *Washington Post*, the *New York Times*, *Newsweek*, *Time*, and TV stations from Japan, Germany, and England.

The proprietor of the Flame and the French Quarter, Bryan West, said it was a cause-and-effect situation. The workers crowded into the bars because they wanted to see the bikini-clad exotic dancers. And because crowds of workers were always there, paying $1.75 per beer, the pimps and prostitutes followed.

The Flame recruited go-go dancers from as far away as New York City, advertising that adventurous women could make a $200 salary, plus $1,300 a week in tips, for dancing in Fairbanks. The New York agent for the Flame interviewed potential dancers in a rundown building in Manhattan.

"The kind of girls we are sending have to be of good character," agent Ralph Clifford told *Harper's Weekly*. "I even originally put the word 'wholesome' into the ad, but I took it out because it seemed to inhibit some of the girls from really dancing. We want girls who want to go there to earn some money, see Alaska, and then get the hell out when the eight weeks are up. Of course, they have to be really strong girls. These men who come down from the oil pipeline, and that's mainly who the customers are, they've been up in the mountains for weeks, some of them haven't seen a woman in who knows how long, and they can be pretty crazy. Our girls have to be prepared to be propositioned every five minutes, but also be prepared to resist."

In Fairbanks, the dancers said the $1,300 in weekly tips was one of Alaska's great myths. Two dancers interviewed in early 1975 said that $20 to $30 a week in tips was more like it. One of the women did say that she was offered twenty drinks one night, however.

In Seattle, nightclub operators and agents complained about a talent shortage. "There has been a serious run on our exotic dancers by loaded Alaskans," said Wally MacDonald, a theatrical agent in Seattle. One of the women from Seattle, a twenty-two-year-old, said working at the Flame was "like being a dance hall gal at the Long Branch Saloon a hundred years ago."

If dancing was a big attraction for bar patrons, so was prostitution. Responding to pressure during one of the periodic outcries against prostitution, West said he had kept prostitutes out of his downtown bars for a month, but he couldn't promise how long that would last. His business was off 50 percent, he told the city council.

One businessman said he counted twenty-seven prostitutes standing outside a bar farther down the street. On the same day a friend of his was asked if he wanted "to party" five times.

The atmosphere downtown was like a circus act. The Lacey Street Theater, nicknamed the "Racy Lacey" by some, had taken to showing films like *Naughty Nymphs* on Second Avenue.

"Downtown theaters have lost the family trade," the president of the theater company said when he visited Fairbanks. The film showing at the Lacey that day was the X-rated *Linda Lovelace for President.*

William Humphries, a well-known local character, said the city ought to buy up all the bars and liquor store licenses downtown to clean up the place.

There were crime problems throughout the day, but most of the fights happened just before the bars closed at 5:00 a.m. "There was always trouble at five o'clock in the morning," said policeman J.B. Carnahan. "It was like a riot that took place every morning."

Some of the worst disturbances took place on an August weekend in 1975. One version of the story is that the battles started because some of the black prostitutes were trying to "cut in" on some of the customers of the Native prostitutes. It wasn't racial strife, it was a turf fight that got out of hand, said a bouncer at the Chena Bar. Others said the brawls had no specific cause, other than the intoxication of the participants.

Fairbanks policeman Victor Gunn was called to the Roustabout at about 4:45 a.m. when a fight broke out. He settled that one and prepared to leave. "I came back outside and the crowd was having at it," he said. There were seventy people throwing beer bottles, rocks, sticks, and stones.

Another policeman on the scene, Lieutenant R.K. Bonneville, said this was different from the usual large crowds that took to the streets at 5:00 a.m. It was worse the next morning, with about one hundred combatants at closing time. About a dozen arrests were made, and several people were stabbed.

Marie Scholle, hired earlier that year as one of the first women officers in the Fairbanks Police Department, said she watched on the second morning as police and Troopers in riot gear prepared to move in. The scene reminded her of "gladiators dressed in armor getting ready to do battle."

There was talk of closing the bars earlier, but nothing came of it.

Carnahan said that at one point, the police realized that the daily fights at closing time could be kept to a minimum if the police pulled back a bit.

"It dawned on most of us that the reason we were having these fights was that we were the object of them. Everybody would bust out of the bars and say, 'Okay, let's fight the police.'"

The police and the Troopers were understaffed during pipeline construction, which made matters worse. Fairbanks Police Chief Bob Sundberg had thirteen unfilled police positions at that time, mainly because the city wages were so low that the department couldn't compete with the pipeline. A month's pay for a policeman was a week's pay on the pipeline. Sundberg eventually joined the exodus too, becoming chief of security for Alyeska.

With the departure of experienced officers, many new recruits signed on, some of them former military MPs and others with little training. The city hired officers who were given badges and guns and were sent out on the street for on-the-job training. It was baptism under fire.

Carnahan said the police were swamped and couldn't keep up. So were the Troopers. They didn't spend much time writing traffic tickets or going after drunk drivers.

"You didn't have to look for work," said Trooper Jay Yakopatz. "It just gravitated in your direction. You had to get out there fast, because the calls were coming in as you gassed up the patrol vehicle."

Captain Jim Vaden of the Troopers said it seemed Fairbanks had more crime per capita than the rest of the state put together. "There were a lot of assaults on police officers. It almost seemed that anytime an officer walked into a bar, someone would pull a knife or haul off and try and brain him with a bottle for no reason at all," he said. "It was unbelievable."

As bad as crime was, however, the problems did not affect the majority of the citizens of Fairbanks. After all, most of them didn't go downtown with $5,000 in their pockets or spend every night in the bars getting drunk.

"The city was always safe," Carnahan said. "This was an activity between people who chose to play. For them it was serious. A lot of them didn't make it through it, but it was of their own choosing. There were no real victims in the pure sense of the word.

"I don't recall anybody wandering down there selling Bibles who got bumped on the head and robbed."

District Attorney Harry Davis, appointed to his job in 1975 at age twenty-seven, agreed that the entire community was not under siege: "The reputation Fairbanks had Outside was that you could get shot on the streets. That really wasn't the situation. Most of the crime was pretty much limited to the

An appeal from some audience members greeted President Gerald Ford in December 1975 when he visited Eielson Air Force Base.

night people, the bar crowd."

In one sense, however, Fairbanks deserved its reputation as a Wild West crime capital, in Davis's view. The police, the Troopers, the prosecutors, and everyone in law enforcement worked in a crisis mode for months at a time, he said.

The combination of pimps, prostitutes, drugs, drunk construction workers laden with bulging wallets, and others intent on nonstop partying, created big-city problems downtown. Officers remember the shootout in front of the King's Kup between warring pimps with automatic weapons, the difficulty of breaking up brawling construction workers, and trying to keep the peace.

Davis said many minor crimes were going unreported and uninvestigated. The Troopers wouldn't respond to many burglaries because they didn't have the staff.

In September 1976, Governor Jay Hammond appeared on KUAC-TV in Fairbanks at the end of an eleven-day stay in Fairbanks and commented on how the police were unable to contain the situation on Second Avenue.

"The environment of downtown Fairbanks has vividly

changed lately and in a way, I think, most Fairbanks residents do not appreciate," the governor said.

Hammond said there was a "crisis in law enforcement," a comment that raised the hackles of the mayor, the city manager, the police chief, and others in the power structure who claimed that everything was under control. A couple of weeks later Hammond apologized to those affronted by his remarks, but in truth things got out of hand during the pipeline boom.

The usual victims were the kind of people Harry McNeal saw downtown in one of the hour-long lines waiting to cash a check. "I've stood in the bank absolutely appalled to see men fresh off the line cashing up to nine of the legendary pipeline paychecks and putting the proceeds in cash into their wallets," McNeal said. "It's grossly unfair to our local Smokies to expect them to protect people from the results of such foolhardiness."

One man fresh from the Slope had about $3,000 in his pockets when he began a night of celebration at the Northward Lounge. To mark his last day in Fairbanks, he was buying drinks for friends and tossing about $100 bills as if they were coupons clipped from the newspaper.

He woke up in the Fairbanks Hotel with $300 in his pockets and told police that he had only spent $500 to $600 on drinks, so the rest must have been stolen when he was incapacitated by drink, which caused him to miss his flight to Seattle.

On Second Avenue, in the words of one observer, "every night was Saturday night, and Saturday night was like New Year's Eve."

Still, it was a tale of two streets. "In the daytime you could go down Second Avenue, you would never believe that it was the same Second Avenue as at night," Scholle said. "It was as if at five o'clock somebody turned the dial to a new channel.

"The politicians saw that we were going to have a problem, but they failed to do anything about it until it was here. That's the way the whole history of Fairbanks has been. It was crisis management."

Pipeliners spent a bundle at the jewelry stores, carried thousands in cash, wore expensive gold nugget watches, and many chose to drink themselves unconscious, which made them easy marks.

"There were guys who got rolled by the same gal, three, four, and five times, and would report it to the police, that's what was always amazing to me," Carnahan said.

A lawyer who had his office on Second Avenue at that time says that pimps would come into his office and count out $1,000 or $1,500 in cash to get him to represent the prostitutes under their control. In many cases, prostitutes were out on bail before the paperwork for their arrest was completed.

There was also the local gambler whose establishment was visited by some intimidating individuals. The gambler thought he was going to be killed by operators moving in on the action.

When the intimidating individuals identified themselves as police officers, the gambler said "Thank God" as he was arrested.

Carnahan remembers walking into a bar just as a woman shot her boyfriend. He also remembers the story of the pipeliner with a big wad of cash on the bar who died at the bar. There were men sitting on each side of him and they kept drinking, using the dead man's cash until it was gone.

"It was a good thing to live through all that," Carnahan said. "I don't think I'd want to do it again."

Oh Alyeska, Oh Alyeska,
Where are your lovely yellow trucks?
They're always at the Elbow Room,
through summer sun and winter gloom.
Oh Alyeska, Oh Alyeska,
Where are your lovely yellow trucks?

— **Sung to the tune of O *Christmas Tree* at
the Alyeska Christmas party in Fairbanks,
December 1974**

THAT SHADE OF YELLOW

The Alyeska Pipeline Service Company trucks, all of them brand new and all of them bright yellow, couldn't have been more visible if they had all been equipped with flashing lights.

"In a town where people drove dusty old trucks that were about to require the parts of the salvaged vehicles rusting in their backyards, the spiffy Alyeska fleet stood out on the roads like so many gold nuggets in a time-worn stream bed," wrote Mim Dixon, director of the Fairbanks North Star Borough Impact Information Center.

Displaying the type of thinking that Henry Ford had used when he made the Model T available in "any color, as long as it is black," Alyeska chose a color scheme that would stand out in the snow. It led to new variation of an old joke: "Don't eat yellow snow. It may be an Alyeska truck."

In July 1974, Alyeska placed a $3 million order at Tip Top Chevrolet in Fairbanks. With that one order, the company bought 450 three-quarter-ton, four-wheel-drive Chevy trucks, as well as 97 Suburbans, 30 Blazers, and six ambulances. It also bought a substantial number of Chevies from Alaska Sales & Service in Anchorage.

"In the summer of '74, there were so many Chevrolet trucks around here [Fairbanks], you couldn't believe it, " said Fred Gray, a Chevy representative on the pipeline project. "Fairbanks was full of Chevy trucks—10 through 30 series pickups, Blazers, Suburbans, all four-wheel-drive units."

Fairbanks Daily News-Miner

Alyeska's fleet of yellow trucks became lightening rods for angry Fairbanksans who'd grown weary of the strain on their city's roadways and services.

Across the state there were 1,700 Chevy trucks on the project. *Chevy* magazine included an article on the use of its trucks, and said, "Trucks wearing Chevrolet bowties have become as familiar as the magnificent scenery and the long work hours."

The Alyeska logo was carried in black letters on the doors of the trucks, along with the words "Exempt Carrier." Some people in Fairbanks took this to mean the company was exempt from fuel taxes, a false rumor that even led to complaints with the borough's Pipeline Impact Information Center.

To those left out of the big money jobs, and to those who were reluctant to see Alaska change, the yellow trucks were the rolling representation of everything they didn't like about the pipeline: oil company influence, hordes of new workers interested only in paychecks, long lines, and traffic congestion.

As the ever-visible symbol of Alyeska in action, the trucks became a public relations nightmare for the company, particularly among those who thought the occupants of the trucks were not working.

"We have been concerned when seeing what appears to be hundreds of those bright yellow trucks and vans running around and seeing numerous Alyeska buses transporting a mere handful of people," said editor Tom Snapp of the *All-Alaska Weekly.*

In the cramped confines of downtown Fairbanks, where the major attempt to accommodate the pipeline traffic boom was to make the major streets one-way thoroughfares, the yellow trucks became a visible irritant to many. An example of the frustration some long-time Alaskans felt, a Fairbanks woman, long active in political circles, complained to Alyeska after a driver in a yellow truck refused to let her change lanes on Noble Street. She was driving her 1948 Buick when she was passed by a yellow truck.

"I didn't appreciate the 'finger' he gave me," the woman wrote later to an Alyeska official. "However, I deserved it because when he refused to allow me into the left lane (so I could exit Noble to the east), I yelled at him as he passed, 'Screw Alyeska!' And I certainly meant it! Several times I've been tempted to go up to a yellow pickup and kick the hell out of it, but that would only hurt my foot, so I've refrained."

The yellow trucks became like a lightning rod for resentment. Dixon said that after Alyeska realized the problem, it shifted many of the yellow trucks to other sites and "ordered new trucks in a variety of more subdued colors."

Trucks, yellow and otherwise, figured prominently in stories told and retold about workers who left the jobsite in company vehicles.

"They had so many pickups that if somebody dropped a pie plate up there, thirty pickups gathered around it right now," trucker Fred Austin said. "Guys would go out on R&R, grab a pickup, drive to the airport, jump in the airplane, and just leave the airport with it sitting there running. They were forever chasing down pickups."

In February 1975, KFAR radio repeated a rumor that three hundred Alyeska trucks were missing. At the time Alyeska had as many as six hundred pickups in the Fairbanks area and a total of about fifteen hundred on the project. The company conducted "Operation AREA," for "Alyeska Rolling Equipment Audit" to find out exactly how many were missing.

The *Los Angeles Times* quoted one pipeline official, a statement that was later denied, as saying two hundred trucks were missing in Alaska, and they had been "scattered from Miami to Mexico City." Alyeska said a more accurate figure was twenty-five to thirty lost trucks.

To try and ease the strain over the Alyeska trucks that stuck out like yellow thumbs in the traffic tie-ups of downtown Fairbanks, the company ordered project vehicles to stay away from downtown when the carbon monoxide level was elevated.

Both in Fairbanks and elsewhere, it also prohibited drivers with Alyeska vehicles from parking outside a bar or any establishment that served alcoholic beverages. This became a point of contention at Livengood, where the only establishment was Sam's, and an exemption was needed for those who stopped for coffee.

Among kids in 1975 Fairbanks, it became a contest to collect yellow gas caps off the trucks.

"We finally beat that one by stipulating any color but yellow when we ordered more pickups," wrote Alyeska spokesman Larry Carpenter. For the trucks with stolen gas caps, black became the color of choice.

Carpenter kept one of the yellow gas caps as a memento. It was among the items included with his papers when they were given to the archives of the University of Alaska, Fairbanks, after his death.

The yellow truck fear was so pronounced that when, in a mix-up, new city fire trucks arrived painted lemon yellow instead of lime green, there was some explaining to do. "Everyone thinks it's an Alyeska truck," said Lieutenant Ken Weaver.

The problem was complicated by the lack of a city insignia on the side of the yellow fire truck. All the sign painters in Fairbanks were busy with pipeline work when the truck arrived, so it remained a no-name truck for a while. "We really get some strange looks when we're going down the road," Lieutenant Jack Hillman said.

"Dear Sirs:

I have been trying to call you for several weeks on the one phone number that you have listed in the Fairbanks directory. I have never succeeded in dialing more than three digits without getting a busy signal and have decided that a letter might be faster."

> — From a January 3, 1975, letter to the Alyeska Pipeline Service Company by Alan Batten of Fairbanks, who wanted to know if trees cut on the Steese Highway were available as firewood.

THE BUSY SOUND OF SUCCESS

CBS newsman Mike Wallace visited Fairbanks for a *60 Minutes* report on how "the panic over energy in the Lower 48 dictated not only that the pipeline should be built, but built fast, and hang the expense."

He said those who didn't believe all the stories about the fortunes to be made on the pipeline could find all the proof they needed in the lobby of the Alaska National Bank of the North, an institution presided over by Frank Murkowski, a future U.S. senator.

"Here's where those pipeline checks turn to Alaskan green," Wallace said from the bank lobby. The camera flashed on a teller counting out $1,923 for a pipeliner. "And while the tellers are in their cages, counting out the money, banker Murkowski smiles a lot," Wallace said.

Much to the dismay of business leaders in Fairbanks, Wallace's report focused entirely on the turmoil in the city. This was in keeping, however, with the general trend of pipeline news coverage.

The *60 Minutes* segment was broadcast on August 17, 1975.

"A visitor who knew the old Fairbanks first notices the traffic," Wallace told his audience that Sunday evening. "There didn't used to be any. Now, if you want to stop downtown, there's just no place to park. The land of muktuk and caribou steak engulfed by fast-fried chicken and cheeseburgers. And communications? Forget it."

To illustrate the communications problem, Wallace simply picked up a telephone.

"We thought perhaps you'd like to hear a boomtown melody," Wallace said. "Folks here call it the song of Fairbanks."

Those who heard the song day in and day out were desperate for a new tune. They wanted to hear that old dial tone.

In a letter to the city-owned phone utility run by Municipal Utilities System, one resident talked about the day he picked up the phone at 9:19 a.m. to make a local call.

"Twenty-six minutes later I got a dial tone, beating your previous record of twenty-three minutes to get a dial tone," said Brian Rogers, a future state legislator and university vice president. "Perhaps if you really work on it, you can get the Fairbanks telephones set up so that no one gets a dial tone at all, thus saving us all the aggravation of having to use the telephone."

The phone frustration factor hit everyone, whether it was a delay of the move of state offices because of a shortage of phone lines, or incessant busy signals that tied up communications every day.

One phone user who couldn't get service took a hammer to a public phone in the utility's office downtown, an extreme example of the exasperation thousands came to live with.

"People were patient several years back," chief phone operator Rose Messina said in 1975. "That patience is gone. If a person has a problem in dialing, we will assist if it's an emergency or an urgent call. Other than that, they just have to keep trying, because all circuits are usually busy. It's our only problem."

Messina said the biggest headache for the operators was the high volume of calls from the Lower 48 with people looking for jobs or trying to call people working in pipeline camps and having no knowledge where they were.

Fairbanksan Jim Hunter, who had been an outspoken opponent of the pipeline, knew about this first-hand. He had a phone number that was one digit off from the number for the pipeline project in mid-1974. Day and night, Hunter was getting calls from people looking for jobs. Fed up, he decided that until the numbers were changed, he would answer his phone as "TAPS Switchboard" or "Bechtel," and "do the best of my ability to provide misinformation."

"In this way I wish to wake the sleeping giant that the problem is not only mine but his also," he wrote.

The sleeping giant's main switchboard at Fort Wainwright soon had a new number, a change that had already been in the works.

No matter what the phone numbers, the system was so overloaded that it was not easy to let your fingers do the walking. A salesman who had been asked to keep track of his calls reported that of 197 attempts, he got through four times. Another caller claimed that he dialed eighty-seven times before completing a local call. The odds of winning may have been better in Las Vegas.

While the system was an improvement over the 1940s, when people would "call" cabs by turning on the porch light, the population increase brought by the pipeline was almost a knockout punch. If it was a local call, you could often drive to the destination and deliver the message in person before getting through on the phone.

Jerry Bowkett, the spokesman for the University of Alaska, told the media in Fairbanks that he would drop off news releases in person: "We are finding it almost impossible to reach you by telephone and know that you are experiencing the same degree of difficulty, too, attempting to contact your various news sources."

To overcome the phone hang-ups, some in the news media and numerous businesses turned to communication by telegram and telexed messages. Tom Snapp, the editor and publisher of the *All-Alaska Weekly*, complained about the phones in a telegram to the Anchorage office of RCA, the long-distance company, and also used it as an editorial: "When we couldn't get through to you by phone, we attempted to reach RCA office locally to send you a telegram. But we could not do so. We spent hours dialing, only to receive a busy signal. We have decided we will hand deliver this telegram to the local the RCA office. So far as we are concerned, the telephone

system, both locally and statewide, is inoperative, unusable."
Aside from long delays for a dial tone, there were other problems as well. Once callers had a dial tone and started to dial,
they often got a busy signal before finishing the call. Even with
a dial tone, callers couldn't be sure that the phone was ringing
on the other end or that it was ringing where they expected it to
be. On some occasions, they would complete the call only to
end up in the middle of someone else's conversation.

Fairbanks journalist Jane Pender told about the time she
dialed two digits and found herself listening to a conversation
between two hotel clerks trying to find rooms for extra
customers. She said she listened and "found out who will
double up and who has to have a room to himself, and what
it's like to be a hotel clerk in what they called our 'combat
zone.'"

Resident James A. Smith joked that this phenomenon was
caused by a gremlin named "Watergate Willy," who "cuts
you in on some of the most interesting two-way conversations.
The other parties don't know you are there so it's just like
wiretapping."

The pipeline didn't create this problem or any of the others
that plagued the phone system. Instead it was the city's failure
to provide money for modernization that was to blame.

"In the telephone business any day you don't expand you
pay for it in the future," said phone manager Earl Land. "And
there's been many days in the past that this organization here
did not expand. . . . There was time lost all along in this organization for years and years."

In 1973, the city started spending money to try to catch
up, spending about $14 million trying to increase capability
in a five-year span. But the increases couldn't come fast
enough to improve traffic flow on a phone system as clogged
as a Los Angeles freeway. The phone company advised
Fairbanksans: "Avoid lengthy and idle conversation. Know
what you're going to say before you place your call. Wait for a
dial tone before you dial."

Under such conditions, Fairbanks seemed to some people
to be one of the most inaccessible places on earth.

Frank Walczyk's mother suffered a stroke in Colorado in

With oil flowing through the completed pipeline in 1977, a graffiti writer added his sentiments to the pipeline near Goldstream crossing.

January 1976. His family in the Lower 48 tried to call him for four days, but was unable to get through to his house, served by the Glacier State Telephone Company. They finally reached him via an emergency police message and told him that she had died.

"If we had decent, competent phone service, my husband could have flown home and seen his mother alive," Walczyk's wife said.

The Fairbanks phone system was "Impact Crisis No. 1" in the Interior city during pipeline construction. In a survey in 1976, more than two-thirds of those polled said the phone service was somewhere between "poor" and the "worst you could expect." Some relief came with the installation in the summer of 1976 of a new $3.5 million computerized telephone switch. One measure of how eagerly the improvement was awaited can be found in the suggestion by Bob Hanson, chairman of the Public Utilities Board, that $5,000 be included in the budget for a citywide beer bust.

The city authorities had hoped to make the first call on the new system to President Gerald Ford at the White House. MUS General Manager Wally Droz said, however, that the president was out of the country and Vice President Nelson

Rockefeller also was out-of-reach. "The vice president isn't too willing to take a call at two in the morning," Droz said.

So the city settled for a first call to U.S. Senator Ted Stevens, placed by former utilities board member Don Chandler and broadcast on local television.

In the days after the new system was installed, it was deluged with calls. MUS Phone Manager Earl Land said they had thought they were uncorking a champagne bottle, but it was really a fifty-five-gallon drum. About 400,000 calls were attempted on the first business day, whereas only 135,000 attempts had been made on a typical day with the old equipment.

With the new telephone switch, MUS configured the system so that callers had to dial all seven numbers instead of just the last five. This provoked traditionalists who claimed that Fairbanks didn't have to be just like everywhere else, but most people were happy to dial all seven numbers if it meant getting a connection.

After the first month, however, lawyer Andrew Kleinfeld still didn't see much of an improvement. "My experience has been that the telephones are just as bad now as they were before the change. It still takes about three tries to make a call."

PLAIN GREED

It was a remarkable display of brevity from the lawyers in the Tanana Valley Bar Association. The resolution approved by the group in November 1975 contained just four words: "Don Gilmer was right."

Gilmer was the Fairbanks North Star Borough planning director who had resigned the previous month and given a remarkable farewell speech in which he said Fairbanksans were becoming greedy.

"I am concerned about the greed that this community is now showing and that greed is probably worse in the people who have been here the longest," Gilmer said to a pro-business group at its weekly breakfast meeting. "It's not the pipeliners, it's not the newcomers, it's the people who have been here five, ten, fifteen, twenty years.

"There are prices being charged here now that have no reason to be charged except for the lack of competition," he said.

Gilmer complained that instead of raising prices by two cents, the stores jacked them up by twenty cents. One game in his family was to find items with different prices. The winner was a box of cake mix at a store with four different prices on it, representing a sixty-five-cent difference.

As to the resolution that said, "Don Gilmer was right," it was an effort by the lawyers to needle the Chamber of Commerce, which had opposed holding the state bar convention at the university. Instead, members of the chamber wanted to spread out the visiting attorneys in downtown hotels and bus them to a central site.

SECTION VI

THE AFTERMATH

Tricia Brown

Today at Prudhoe Bay, caribou sightings are common and herd numbers remain at healthy levels.

"Over time, each great project has been succeeded by one greater, most of them embodying a leap of engineering vision: irrigation canals, aqueducts, Greek columns, the Roman arch, the Taj Mahal, De Lessep's canal, Eiffel's Tower, the skyscraper, the geodesic dome, the Golden Gate Bridge, the Alaska pipeline, the Chunnel, the Three Gorges Dam.

There are other aftereffects. Frustrated pharaohs, popes, presidents and every homeowner who ever decided to add a bathroom bear testimony across history that few buildings have cost what the architect said they would, or were ready when the contractor promised."

— **Los Angeles Times report on the history of big engineering projects, 1995**

BIGGER AND BUSIER

Twenty years after the pipeline went into operation, Alyeska decided to move hundreds of workers and contract employees to Fairbanks from Anchorage. This happened as part of overall cutbacks in the work force, a response to a continued decline in oil production.

The question arose within Alyeska about why these workers were in Anchorage, hundreds of miles from the pipeline route, instead of closer to the action.

The reasoning that had led the company to base much of its operational staff in Anchorage immediately after pipeline construction no longer seemed to apply. Fairbanks business leaders had made a pitch to attract the operations staff of Alyeska in the mid-1970s, but Anchorage had public and private amenities that overrode the advantages of geography.

In the decades after the pipeline ended, the Fairbanks population grew by tens of thousands, and new roads, stores, and subdivisions relieved the growing pains. The pipeliners who performed on Second Avenue in the summer of 1975 would

With warm oil flowing through the pipe in 1977, visitors stopped for a close-up look near Fox, where the pipeline parallels the Steese Highway.

barely recognize the place today. Most of the pipeline-era bars are gone. The city bought the downtown property and demolished the buildings to make room for a hotel that never quite came together. Today there are parking lots and a park with a statue. In summer, the Lacey Street Theater contains refrigerated displays of ice sculpture, and the University of Alaska has a major presence downtown.

Everything in Fairbanks got bigger and busier, far surpassing the peak days of pipeline impact. But because the community was larger, with three major retail centers away from downtown, the pressure cooker atmosphere evaporated.

The start of pipeline operations quickly catapaulted Valdez into one of the country's major ports, in terms of the tons of cargo that flowed away in the bellies of the tankers. In the prosperous years that followed, taxes on Alyeska property allowed Valdez to diversify its economy and to build a grain terminal (for grain that never came) and other facilities, from a museum to a library.

The town boomed again when the *Exxon Valdez* ran aground in 1989, an environmental catastrophe that also drew thousands of job-seekers to Valdez to join the clean-up crews. In the summer of the spill, it was like the mid-1970s all over again in Valdez.

"That single event in 1989 could have a major impact on the world energy balance in the late 1990s to the degree that it tilts the scales against new development in the United States, leading to even higher imports," wrote oil historian Daniel Yergin.

The United States, which now imports a majority of its oil, gets about 10 percent of its daily consumption from Alaska, where oil production has been declining since 1988. At peak production, Alaska supplied about one-eighth of the oil consumed in the nation.

To celebrate the bicentennial in 1976, *Life* magazine published a special edition on the "100 Events That Shaped America."

The editors selected the construction of the Trans-Alaska Pipeline as one of the top hundred events in the history of the nation. If a magazine set out to compile such a list today, I doubt the pipeline would be in the top hundred, which reveals something about our attention span and how events in the news shape our perceptions of history.

In a 1997 discussion on National Public Radio about the importance of "Big Projects" to society, the World Trade Center, the Golden Gate Bridge, and even a far-fetched proposal for a railroad tunnel beneath the Bering Strait were much discussed, but not the Alaska pipeline.

On any list of the 100 Events That Shaped Alaska, the construction of the pipeline would vie with the approval of statehood for Alaska and the Gold Rush as a candidate for the top spot—not so much because of what it did to the landscape and that fabled twelve square miles, now grown to sixteen square miles, but because of the windfall of tens of billions of dollars it brought to Alaska.

Oil money changed Alaska society in every way and left the state so dependent on oil that the economy rose and fell with fluctuations in the world price. Fueled by billions of oil

Tricia Brown

After pipeline construction was finished, all of the camps were broken down and removed as per Alyeska's agreement. But one business owner at Coldfoot decided to buy the used trailers and convert them into sleeping space for visitors. Large busloads of tourists traveling with Gray Line of Alaska or Princess routinely overnight at Arctic Acres Inn, one of the only establishments on the Dalton Highway.

dollars, Alaska took off on a spending spree that brought sports arenas, convention centers, improved utility systems, and other public works projects to towns and cities. And the state amassed a $20 billion savings account for the future, a hedge against the day when the wells run dry.

"I bow down each morning and thank God for the North Slope," Al Adams of Kotzebue once said on the floor of the state House of Representatives.

In the early 1980s, when the money flowed like the Yukon River, the state directed money into everything from opera to little league, offered subsidized loans, gave grants to people for insulating their homes, and sent cash rebates to people who donated to political candidates.

Most Alaskans, on some level, recognized the danger of this heavy dependence on a declining resource, but there were no simple solutions.

As the years went by, Alaskans began to take the wealth produced by oil for granted. More came to accept as normal the idea of living in a society that offered a great deal in the way of government services and little in the way of state taxes, except oil taxes.

As the annual revenue from oil production declined in the 1980s and 1990s, political debate shifted increasingly to whether earnings from the $20 billion Alaska had stashed away should be used to pay for government services, or if the savings account should only be used to continue making annual cash payments of more than $1,000 to every Alaskan. The struggle over public good vs. private gain is one of the lasting images of pipeline impact.

Two decades after the first tanker departed from Valdez, the Alaska economy depends on petroleum the way the Iowa farmer depends on corn. North Slope oil production has declined by hundreds of thousands of barrels per day, but with the development of new oil fields the pipeline is expected to remain operating well into the twenty-first century. The oil in the pipe, flowing across the state at about 45,000 gallons a minute, remains the lifeblood of the Alaska economy.

Tricia Brown

Outside Arctic Acres Inn at Coldfoot stands the original sign that once welcomed pipeline crew members to Coldfoot camp.

Tricia Brown

The Dalton Highway sees little private traffic today, but long-haul truckers continue to do business with oil companies at Prudhoe Bay.

The industry centered on the North Slope has become part of the everyday routine of life in Alaska. Thousands of workers commute to their jobs at Prudhoe Bay, Valdez, and to the pump stations through the darkness of winter and the all-night light of summer.

Most Alaskans give little thought to the invisible river of warm oil that flows south at five to six miles per hour, twenty-four hours a day. It enters the pipeline at a temperature of about 116 degrees and with the heat of friction, drops only to about 92 degrees by the time it comes out the other end. Because the end product is out of sight unless equipment breaks or someone makes a mistake, it's often out of mind.

Just as the electric and telephone lines to your house tend to be ignored as long as the power is on and the phone rings, so it has been with the pipeline. It has attracted intense public attention and controversy only when things have gone wrong.

When I walk from my house in a subdivision north of Fairbanks and see the pipeline, it seems as still as the old rusting gold dredge not far away that hasn't scooped any gravel in decades. Like the gold dredge, the pipeline is a tourist attraction. In the summer tourists stop by the busload to get their pictures taken and to touch the sheet metal covering, beneath which a four-inch blanket of insulation covers the half-inch walls of the forty-eight-inch diameter pipe.

While the dredge is a relic of an earlier day, the pipeline represents modern wealth as surely as there are Cadillacs in Kuwait. At the pipeline turnout in Fox near my house, as in places throughout Alaska, the legacy of the oil pipeline boom is evident, though the state still wrestles with what oil hath wrought.

ALMANAC

Fairbanksans could actually see pipeline construction in spring 1976, when pipe was installed just north of town near Fox. Today a visitor information site near the Steese Highway attracts thousands of tourists each year.

BY THE NUMBERS

Pipeline cost: $8 billion. Because of inflation, cost if built in 1994, $20 billion to $22 billion.

Time to complete: 3 years, 2 months.

Contractors and subcontractors: 2,000.

Airfields: 11 temporary airstrips; 3 permanent airfields at Five Mile, Galbraith, and Prospect.

Construction camps: 29. Largest camp, Marine Terminal, 3,480 beds. Largest pipeline camp, Isabel Pass, 1,652 beds.

Number of cubic yards of earth moved: 93 million. Number of cubic yards of excavation to build the Suez Canal, 98 million.

Government permits required: 515 federal; 832 state.

Elevated pipe: Supported by H-shaped steel supports made of two vertical support members, a crossbeam, and a pipe "shoe." Some 78,000 vertical support members used. 61,000 fitted with heat pipes to keep permafrost below pipe frozen.

Minimum amount of earth covering buried pipe: 3 feet.

Styrofoam insulation panels below gravel on the work pad: 100 million board-feet.

Truckloads of pipe: 19,000.

Air cargo: Largest private airlift in history, transporting more than 150,000 tons to camps north of the Yukon River in 1975.

Spare parts: $175 million worth, enough to strain the supply capabilities of companies like Caterpillar.

River and stream crossings: 34 major; 800 others.

Highest point on the pipeline: Atigun Pass, in the Brooks Range, at 4,739 feet.

Ownership of Alyeska by percentage: British Petroleum, 50.01; ARCO, 21.35; Exxon, 20.34; Mobil, 4.08; Amerada Hess, 1.50; Phillips, 1.36; Unocal, 1.36.

First Tanker: ARCO Juneau departed Valdez on August 1, 1977.

Area covered by pipeline system: 16.3 miles.

Length: 800.302 miles.

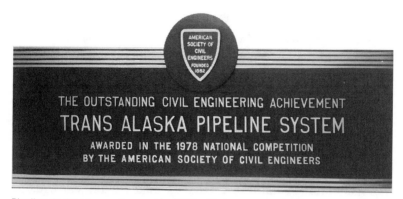

THE OUTSTANDING CIVIL ENGINEERING ACHIEVEMENT
TRANS ALASKA PIPELINE SYSTEM
AWARDED IN THE 1978 NATIONAL COMPETITION
BY THE AMERICAN SOCIETY OF CIVIL ENGINEERS

Pipeline designers were honored in 1978 with the country's highest civil engineering award.

Average depth of Prudhoe Bay oil formation: 9,000 feet.

Cost of Yukon River Bridge: $30 million.

Pipeline animal crossings built: Approximately 579.

Area revegetated through 1993: Almost 7,750 acres.

Pipe thickness: One-half inch.

Weight: 235 pounds per lineal foot.

Amount of pipe: 120 shiploads from Japan. Six shiploads for Atigun Floodplain Pipe Replacement Project from Italy.

Testing: All pipe tested with water pressure tests at 125 percent of normal operating pressure.

Work force: Peak employment 28,072 in October 1975. Total employment from 1969-77, about 70,000.

Minority hire: 14 percent to 19 percent.

Women workers: 5 percent to 10 percent.

Bridges: 13 for the pipeline; 20 on Haul Road, the road that later would be named the Dalton Highway.

Pump stations with refrigerated foundations: 5.

"We're not building a project up there to leak, and we're not building a project up there that won't deliver the oil. That's the main thing to remember. No one in his right mind is going to put almost $8 billion into a pipeline to have it fail."

— U.S. Senator Ted Stevens at a hearing of the Senate Interior Committee

Valdez terminal: Can hold 9.18 million barrels of oil.

Concrete weights to hold down pipe in river crossings: 75,000 pounds for 40-foot section.

Above-ground pipe: 420 miles.

Below-ground pipe: 380 miles.

Time oil will stay pumpable in winter if shutdown: At least 21 days.

Worst accident: *Exxon Valdez* spilled nearly 11 million gallons in Prince William Sound in 1989.

Worst accident on land: 1978 sabotage attack that spilled 670,000 gallons near Fairbanks.

Grounding rods: Atigun pipe replacement has zinc ribbons that act like grounding rods and prevent pipe corrosion.

Insulation: 3.75 inches on elevated pipe.

Maximum capacity: Full pipeline can contain 9,065,065 barrels of oil.

Expansion: 40-foot sections of pipe expand or contract .031 inches with every increase or decrease of 10 degrees.

Zig-zag configuration: Allows pipe to shift as it expands or contracts with temperature changes. Also allows pipe to move during earthquakes. An 1,800-foot section of elevated pipe could vary in length by as much as 18 inches because of temperature extremes of 70 below and oil as hot as 145 degrees.

Average speed of oil from Prudhoe Bay to Valdez: 5–6 mph.

Draft of biggest tankers at Valdez: 85 feet.

Average time to load tankers: 18 hours.

Average number of tankers per month: 58.

Welds: 108,000 required. Cost of welding repair program in 1976 caused by controversy over X-ray falsification and faulty welds, $80 million to $100 million.

Haul Road: Cost $125 million. 358 miles long. 32 million cubic yards of gravel. Built in 154 days. Renamed the Dalton Highway.

Materials: 3 million tons shipped to Alaska; 73 million cubic yards of gravel used.

— *From "Trans-Alaska Pipeline Facts," published by Alyeska Pipeline Service Company, and other sources.*

Bulldozers were modified to carry sections of forty-eight-inch pipe. Here a Cat crawls up to the top of Keystone Canyon, about eighteen miles east of Valdez, in August 1976. The canyon was one of the more challenging engineering and construction areas.

PIPELINE CRACKS

Atigun Camp—The "Official Rumor of the Day Report" said the Rumor Control Board had learned that MGM Studios was going to make a movie about the pipeline and auditions were going to be held. The following parts were to be filled:

Construction Manager: Must have a sense of humor, charm, degree in sociology, and be a good prevaricator.

Assistant Construction Manager: Prefer someone with striking looks, well-groomed, sociable, and be able to sleep with his eyes open.

General Superintendent: Need someone tall, dark, handsome, with Southern accent. Able to bluff Quality Control, construction managers, etc., and never change his expression. (Poker face.)

Foreman (Need 4): Must be able to do all of the above, plus stay on the set longer. (Must join Actors Guild.)

Engineer: Doesn't have to know how to do anything. Just try to look intelligent. (Make-up will be furnished.)

Quality Control (Need 150): Must be able to stand around, try to stay out of the way. Must be tri-lingual.

Teamsters (Need 20,000 or more): Bring your own vehicles. Drive up and down set. Look busy. Do not overdo it.

Welders (Need 4, also 47 helpers): With hats. Do not necessarily have to do anything, but sit on the bus. (Buses will be furnished.) Also, must be able to wobble.

Laborers (Need 100): Bring your own shovels. Take turns sitting on the bus.

Operators (Need 6, also 1 mechanic per operator): Must be able to sit on a sideboom at 40 below and not freeze up. (Little action required for this part—just stamina.)

Prudhoe Bay—Among the most widely circulated jokes was the one about what would happen when the pipeline would start operation. A crowd would gather at Prudhoe Bay. An executive would hit the switch. And the Valdez harbor would be sucked dry.

Tonsina—Pipeline worker Jo Elasanga expressed the longing that the workers felt in a poem: "The thing of which I dream the most, When I have time to sorta coast, Is not of all these worldly things; Like money, cars and diamond rings, but of the day I get in my car, and leave this camp on R&R."

Fairbanks—On any list of creative housing solutions in Alaska, the following deserves a prominent place. This ad appeared numerous times in the *"Unfurnished Apartments"* section of the Fairbanks Daily News-Miner classified ads: *"Looking for a reasonable alternative to high-priced housing and rental? Advance to the frontier living of '75, with your very own home providing all the comforts of high cost living at a price you can afford. Contact Alaska Tent and Tarp for our solution to the housing problem. Cozy white canvas home available in sizes from 6-by-8 feet to 16-by-24 feet, ready to move into. All* *you need is a couple of trees with a little ground between them. Talk to our housing engineers at Alaska Tent and Tarp, 529 Front St."*

Pump Station No. 5—After a fire destroyed the outhouse here, a crew built a new four-holer and a dedication ceremony was held. A toilet paper ribbon was cut and speakers went on to give credit where credit was due.

To Fluor engineering for engineering expertise and unique design.

To the Teamsters for their first contribution.

To the Ironworkers for rigging the delicate contents.

To the Operating Engineers for knowing how to operate the facility.

To the Laborers for cleanup services.

To NANA Security for protection.

To the Carpenters for assisting in removing splinters.

To the Surveyors for proper alignment.

To the Sheet Metal workers for venting.

To the Insulators for seeing to it that important areas don't get frostbitten.

Galbraith Lake—From a posted memo titled *"General Health."* It read: *"Regular bathing is a matter of common politeness. No matter how attractively a meal is presented, a person's appetite may very well be ruined if an unwashed person sits close by. There is no reason to avoid the regular shower while in camp."*

Valdez—On Thursday, July 28, 1977, at 11:02 p.m., oil reached Valdez from Prudhoe Bay, making Jean Mahoney $30,000 richer. She won the first and last *"Great Alaskan Pipeline Classic,"* having purchased a ticket to guess how many days, hours, and minutes it would take for the first barrel of oil to get from Prudhoe Bay to Valdez. She was one minute off from the correct time of 38 days, 12 hours, and 56 minutes.

Fairbanks—At the height of conflict between old-time Alaskans and pipeliners from southern parts, a woman drinking at the

Alyeska Pipeline Service Company

Station No. 5 got an unexpected visitor in August 1976. Poor handling of food was largely responsible for creating nuisance bears.

Boatel Bar along Airport Way went outside to find her car blocked by a blue Buick with Texas plates. When no one moved the car, she got into her own and began ramming it out of the way. "As she drove away, a bewildered Buick owner stood on the steps, slack-jawed in disbelief," a reporter wrote.

Juneau—On April Fool's Day 1976, the Alaska House of Representatives offered to make a trade with the Arab oil barons for nonresident pipeline workers.

The resolution said that because of the overabundance of nonresident pipeliners, the state should offer two of them for every falcon the Arabs would give back to Alaska. The spoof followed the revelation that the CIA had obtained six endangered peregrine falcons in Alaska to curry favor with Arab oil sheiks.

Dietrich—Pipeliner mechanic David McCracken told about the Dietrich man who got drunk, quit, and was unable to get a flight back to Fairbanks. The operator put a barrel of fuel on the bucket of a Bobcat, a miniature front-end loader that could go about 5 mph, and headed south. *"Eight hours later he was seen tooling past the entrance to Coldfoot, and I've heard that he made it almost to Old Man,"* McCracken said.

Fairbanks—When asked if Alyeska Pipeline Service Company might donate sixteen portable toilets to the borough, a company spokesman joked: "We'll do it if we get to put our name on them and paint them all yellow."

Washington, D.C.—"Ten miles of the Alaska pipeline are missing," said comedian Mark Russell. "Authorities are determined to find it—if they have to search every pawnshop in the state. The foreman on the job up there put a sign on the bulletin board that said, 'To whoever took the ten miles of pipeline, put it back where you found it, and no questions will be asked.'"

Galbraith Lake—Seen on the camp bulletin board: "First annual Galbraith Lake Belching Contest. July 13, 1974. Girls Invited. Sign up at Mess Hall with Big Jim."

New York—National Lampoon printed an "official souvenir program" of the "Alaska Pipeliners of the American Basketball Association." The mascot of the team was Pipie II, who was better behaved than the eleven previous polar bear mascots. "In the event that Pipie breaks loose, please remain seated. All our players and our police are armed and can maintain security. Do not rush for the exit. There is only one exit in the arena. Do not panic."

Burbank, Calif.—"Apparently there are some defects in the Alaskan pipeline. I understand the president of Exxon just called a Thrifty Drug Store and ordered a seven-hundred mile roll of Mystic Tape." *Johnny Carson, Tonight Show, June 22, 1976.*

"It's nice to know that despite it all, some things do eventually get built."

— The Wall Street Journal
upon completion of the
Trans-Alaska Pipeline in 1977

Fairbanks—A popular joke going around town went like this: "You know how to get twenty-one Texans in a Volkswagen?

Just tell them it's going to Alaska."

Old Man—Seen on the bulletin board: "If we've learned one thing from camp life at Old Man, it's to not work between meals."

Eielson Air Force Base—Joking during a speech at the base, pipeline chief engineer Frank Moolin, Jr., said: "One Teamster from Galbraith told me he doesn't really consider himself drunk in Fairbanks if he can lie on the floor of the Savoy Bar without hanging on."

Fairbanks—Jack Perkins of the *NBC Nightly News* stood next to a 150-foot test section of pipeline outside of Fairbanks for a news report. This was before any pipe had been laid. A sixth-grade girl brought to the site asked, "Is there any pipe laid south of here?" No. "Is there any laid north of here?" No. "You fellows sure have a lot of pipe left to lay, don't you?" she said.

Fairbanks Daily News-Miner

The first barrel of North Slope crude made its way south well before the pipeline was built. In late September 1969, Alaska Governor Keith Miller took part in the ceremony to transfer a symbolic fifty-five-gallon drum, painted gold, to a helicopter at Prudhoe Bay. From there, the drum was delivered to the *S.S. Manhattan*, an ice-breaking oil tanker, and carried south on a 4,500-mile return trip through the Northwest Passage. With Miller are two officers from the *Manhattan*, Roger Steward, left, and Stan Haas. The barrel of oil was taken from ARCO-Humble's 1 Sag River confirmation well.

PIPELINE TIMELINE

March 13, 1968
Atlantic Richfield Company (Arco) and Humble Oil and Refining Company (now Exxon Company, USA) announce discovery well at Prudhoe Bay.

February 7, 1969
Acronym TAPS, from Trans-Alaska Pipeline System, is born when Atlantic Pipe Line, Humble Pipe Line, and BP Oil Corporation (formerly BP Exploration USA, Inc.) agree to begin design and construction of an eight-hundred-mile pipeline. Plans are announced on February 10.

June 6, 1969
TAPS files for federal right-of-way permits.

September 13, 1969
First load of forty-eight-inch pipe arrives in Valdez from Japan.

October 22, 1969
Three original owner companies are joined by Amerada Hess Corporation, Home Pipe Line Company, Mobile Pipe Line Company, Phillips Petroleum Company, and Union Oil Company of California.

April 1970
Environmental groups and others file suit to stop pipeline construction.

August 14, 1970
Owner companies incorporate to form Alyeska Pipeline Service Company.

November 16, 1973
Trans-Alaska Pipeline Authorization Act becomes law.

April 29, 1974
Haul Road construction begins between Prudhoe Bay and the Yukon River.

March 27, 1975
First pipe is laid at Tonsina River.

October 11, 1975
Yukon River Bridge, later named the E.L. Patton Bridge, is completed.

October 26, 1975
The Trans-Alaska Pipeline is 50 percent completed.

May 31, 1977
Final pipeline weld near Pump Station No. 3.

June 20, 1977
First oil flows from Pump Station No. 1.

The pipeline parallels the Richardson Highway about eighty miles south of Fairbanks, where it crosses the Tanana River on a cable suspension bridge, one of two suspension bridges out of thirteen bridges constructed for the pipeline route. Work is still in progress in this 1976 photo.

July 8, 1977
Pump Station No. 8 destroyed by explosion and fire; one person killed; 300 barrels of oil lost.

July 28, 1977
First oil reaches Valdez Marine Terminal at 11:02 p.m.

August 1, 1977
ARCO Juneau leaves Valdez with first tanker load of oil.

January 22, 1980
One billionth barrel arrives in Valdez.

June 19, 1987
Tenth anniversary of the Prudhoe Bay field and the Trans-Alaska Pipeline.

January 14, 1988
Highest daily throughput of 2,145,297 barrels.

March 24, 1989
Exxon Valdez oil spill of 260,000 barrels when vessel runs a ground at Bligh Reef.

1997
Twelfth billion barrel expected to arrive in Valdez.

SOURCES

In addition to interviews with participants, oral history tapes from a private collection of Michael Carey, and a state collection owned by the Alaska State Library, I found information contained in the Larry Carpenter collection at the University of Alaska Fairbanks Rasmuson Library to be particularly valuable. I also consulted the records of the Fairbanks Chamber of Commerce, the Fairbanks Environmental Center, and other archival collections.

Dirk Tordoff of the University of Alaska Fairbanks Rasmuson Library helped me get access to copies of tapes of a series of one-hour broadcasts on KUAC-TV in Fairbanks that featured pipeline executives lecturing a class about the project in the fall of 1975.

Ben Logan, of the Los Angeles office of O'Melveny & Myers LLP, who worked with Alyeska during the prolonged tariff case that went on until the mid-1980s, graciously provided a few thousand pages of testimony and briefs filed with the Alaska Public Utilities Commission and the Federal Energy Regulatory Commission.

The news coverage of the *Anchorage Daily News*, the *Anchorage Times*, *The Pioneer All-Alaska Weekly*, the *Fairbanks Daily News-Miner*, and the *Los Angeles Times* was especially helpful.

I also consulted the news coverage of numerous other organizations including: *The Seattle Times*, the *Seattle Post-Intelligencer*, the *New York Times*, the *Wall Street Journal*, the *Christian Science Monitor*, the *New York Daily News*, the *Washington Post*, *Time*, *Newsweek*, *Forbes*, *Audubon*, *Smithsonian*, *Good Morning America*, *60 Minutes*, *NBC Nightly News*, *Paul Harvey News and Comment*, *Alaska* magazine, *Alaska Advocate*, *Alaska Construction & Oil*, *The*

New Yorker, New Times, the *Chicago Tribune,* and the *Minneapolis Tribune.* Among the variety of documents published by the pipeline builders, articles in the *Campfollower* newspaper and the *Alyeska Reports* magazine were most useful. The following is a partial list of the books consulted:

Allen, Lawrence J. *The Trans-Alaska Pipeline. Vol. 1: The Beginning. Vol. 2: South to Valdez.* Seattle: Scribe Publishing Co., 1975 and 1976.

Baring-Gould, Michael and Bennett, Marsha. *Social Impact of the Trans-Alaska Oil Pipeline Construction in Valdez, Alaska 1974-1975.* Anchorage: University of Alaska Anchorage, 1976.

Brown, Tom. *Oil on Ice: Alaskan Wilderness at the Crossroads.* Edited by Richard Pollack. San Francisco: Sierra Club Battlebook, 1971.

Coates, Peter A. *The Trans-Alaska Pipeline Controversy.* Bethlehem, Pa.: Lehigh University Press, 1991.

Chesterfield, Allen. *The Alaskan Kangaroo: An Odyssey of the Alaskan Pipeline.* Seattle: Adelaide Press, 1980.

Dixon, Mim. *What Happened to Fairbanks? The Effects of the Trans-Alaska Oil Pipeline on the Community of Fairbanks, Alaska.* Social Impact Assessment Series. Boulder, Colo.: Westview Press, 1978.

Dobler, Bruce. *The Last Rush North.* Boston: Little, Brown and Co., 1976.

Fineberg, Richard A. *Pipeline in Peril: A Status Report on the Trans-Alaska Pipeline.* Ester, Alaska: Alaska Forum for Environmental Responsibility, 1996.

Hanrahan, John and Gruenstein, Peter. *Lost Frontier: The Marketing of Alaska.* New York: W.W. Norton, 1977.

Kruse, John A. *Fairbanks Community Survey.* Fairbanks: Institute of Social and Economic Research, 1976.

Lenzner, Terry F. *The Management, Planning and Construction of the Trans-Alaska Pipeline System.* Washington, D.C.: Report to the Alaska Pipeline Commission.

Manning, Harvey. *Cry Crisis! Rehearsal in Alaska (A Case Study Of What Government By Oil Did to Alaska And Does to the Earth).* San Francisco: Friends of the Earth, 1974.

McGinniss, Joe. *Going to Extremes.* New York: Alfred A. Knopf, 1980.

McGrath, Ed. *Inside the Alaska Pipeline.* Millbrae, Calif.: Celestial Arts, 1977.

McPhee, John. *Coming Into the Country.* New York: Farrar, Straus and Giroux, 1976.

Mead, Robert Douglas. *Journeys Down the Line: Building the Trans-Alaska Pipeline.* Garden City, N.Y.: Doubleday & Co., 1978.

Roscow, James P. *800 Miles to Valdez: The Building of the Alaska Pipeline.* Englewood Cliffs, N.J.: Prentice-Hall, Inc., 1977.

Romer, John and Elizabeth. *The Seven Wonders of the World: A History of the Modern Imagination.* New York: Henry Holt and Co., 1995.

Simmons, Diane. *Let the Bastards Freeze in the Dark.* New York: Wyndham Books, 1980.

Strohmeyer, John. *Extreme Conditions: Big Oil and the Transformation of Alaska.* New York: Simon & Schuster, 1993.

Wickware, Potter. *Crazy Money: Nine Months On The Trans-Alaska Pipeline.* New York: Random House, 1979.

Wolf, Donald E. *Big Dams and Other Dreams: The Six Companies Story.* Norman, Okla.: University of Oklahoma Press, 1996.

Yergin, Daniel. *The Prize: The Epic Quest for Oil, Money and Power.* New York: Simon & Schuster, 1991.

Index

A

Alaska Native Land Claims
Settlement Act 20
Alcohol 121, 176–177, 184
All-Alaska Weekly
131, 133, 135, 149
Alyeska Pipeline Service
Company 16, 20, 28, 30,
186–189, 198–200
public image 18
"Sweepstakes" 34–36
yellow trucks 186–189
Anchorage Daily News 143–144
Atigun 70, 80–81
Austin, Fred 41, 44

B

Baer, Wally 157
Bechtel Corporation 36
Bicentennial 112–113
Event that Shaped America
200
Boucher, H.A. "Red" 156
Brooks Range 73
Brower, David 18
Business conditions 126–129

C

Carey, Fabian 152–153
Carlson, John 154
Carnahan, J.B. 181–182, 185
Carpenter, Larry 155–156
Carr, Jesse 40, 44, 142–144
Chandalar 48
Chesterfield, Allen 81
Coldfoot 48, 52, 56, 72, 79–80
Craft, Greig 118–121
Crime 181–185
Croft, Chancy 21

D

Davis, Harry 182–183
Delta 75
Dexter, Raymond 93–94,
100–102
Dietrich 54, 73, 82
Dixon, Mim 146
Drugs 76–77

E

Exxon Valdez 18, 200

F

Fairbanks 11–13, 24
attitudes about development
153–156
post-pipeline 199
prices 196
Fairbanks Daily News-Miner
11–13, 101, 134, 159, 171
Fitzsimmons, Frank 142–143
Fleming, Al 24–25
Ford, Gerald 86–87

G

Galbraith 74, 79, 83
Gambling 74–76, 123
Gann, Ernest K. 87
Generous, Mabel A. 32
Gilmer, Don 196
Gross, Dan 28
Growth 152–165
attitudes about 153–155

H

Hammond, Jay 157–158,
183–184
Haugen, Dave 33, 55, 60–61
Hoppner, Lloyd 174–175
Housing shortage 163, 167–173
Hughes, Shirley 79

J

Journalists 11–13, 21, 24–26

K

Kissinger, Henry 85–86
Knight, Gladys 85–86
Koponen, Niilo 39

L

Labor efficiency 28–31
Laborer's Union Local 942 27
LaRocca, Joe 159
Lenzner, Terry 29
Little, Sam 38–39, 41, 43
Local 798 32, 57–64
Los Angeles Times 133, 148
Lundell, Glenn 50

M

Martin, Jack 142–143, 145
McCasland, Marty 107–108
McCracken, David 94, 150

McDermid, Mike 73
Mead, Robert Douglas
    31, 72, 149
Menges, Phil 42, 115
Merriman, Margaret
    (Muleskinner) 114–116
Miller, Melva 45–47
Money 24–26, 118–129
    financial impact on state
    200–204
    jobs in town 128–130
Moolin, Jr., Frank 12, 21,
    30–31, 33–37, 49, 147

N

News coverage 12, 24–26,
    30, 54, 136, 143–144, 179
North Slope Haul Road 38–44,
    114–116
North Star Terminals Murders
    142–144

O

Oil companies 29–31
Oil impact 11
Oklahomans 73–74
Old Man 82, 103

P

Paskvan, Tommy 178
Patton, Ed 16, 27, 39, 119,
    147–149
Pender, Jane 134
Pettus, Harry 142–143, 145
Pipe 20, 109
Pipeline camps 66–84
    chaplains 100–102
    food 91–98
    newspapers 84
    rules 89–90
    social atmosphere 82
    streakers 99
    TV 103–105
Prostitution 132–141
    Anchorage mayor's
    commission 140
    attitudes in Alaska 132–133
Prudhoe Bay 18
Pump Stations 69–70

R

Rest & Relaxation 52, 79,
    124–126

Roche, Jud 53–54

S

Second Avenue 132–136,
    176–185
    concentration of bars 176–177
Sexual pressures in pipeline
    camps 51–56
Snapp, Tom
    131, 133, 135, 143, 158,
    192–193
Souvenirs 109–111
Stabenow, Dana 60

T

Tallman, Barbara 52–53
Teamsters Union Local 959
    40–44, 62–63, 118–119
    attitudes about news coverage
    143
    union's decline 145
Telephones 84, 165, 190–195
Theft 146–150
Tonsina 57, 62–63, 97–98
Traffic 156, 174–175
Trans-Alaska Pipeline 15–22
    celebrity visitors 85–87
    cost 21, 28
    environmental debates 19–20
    One of Seven Wonders of U.S.
    15
    route 15–16

U

Unions 26–31, 142–144
    attitudes toward development
    165–166

V

Voigt, Rudy 39–40

W

Welders 46, 57–64, 97–98
Weschenfelder, Suzy 45, 51
Wickware, Potter 28, 59, 110
Wolfe, Richard 132
Women 45–56, 78–79, 82
    adjustment to life on line
    51–56
Workers 24–32, 101–102

Sam Harrel

About the Author

Dermot Cole is a long-time newspaper columnist for the *Fairbanks Daily News-Miner.* Cole grew up in Pennsylvania and lived in Taiwan, Hong Kong, and Montana before moving to Alaska at the start of the pipeline boom. He studied journalism at the University of Alaska Fairbanks and was named a Michigan Journalism Fellow in 1986-87 at the University of Michigan. He also worked for the Associated Press in Seattle.

He lives north of Fairbanks with his wife, journalist Debbie Carter, and their children, Connor, Aileen, and Anne. They enjoy cross-country skiing in winter and camping, soccer, and softball in the summer.

Cole is the author of two other books: *Frank Barr, Bush Pilot in Alaska and The Yukon* and *Hard Driving: The 1908 Auto Race from New York to Paris.*